Geography and Geographers

Anglo-American
Human Geography since 1945

DATE DUE

Geography and Geographers

**Anglo-American
Human Geography since 1945**

R. J. Johnston
Professor of Geography, University of Sheffield

Edward Arnold

© R. J. Johnston 1979

First published 1979 by
Edward Arnold (Publishers) Ltd
41 Bedford Square, London WC1B 3DQ
Reprinted 1981 with additions to the Bibliography

British Library Cataloguing in Publication Data

Johnston, Ronald John
  Geography and geographers.
  1. Anthropo-geography – Great Britain
  – History – 20th century
  2. Anthropo-geography – United States
  – History – 20th century
  909   GF28.G7

  ISBN 0-7131-6239-2
  ISBN 0-7131-6238-4 Pbk

Photoset by D. P. Media Limited, Hitchin, Hertfordshire
Printed by Spottiswoode Ballantyne Ltd., Colchester and London

Contents

A Note to the Reader

This book comprises a book within a book. The inner volume—Chapters 2–6—contains the factual record of the history of Anglo-American human geography. The outer volume, which incorporates Chapter 2–6 with Chapters 1 to 7, is an attempt to account for that history. Thus the outer volume is the 'innovative' work, based on the material of the inner volume: Chapter 7 interprets Chapters 2–6 in the light of the model presented in Chapter 1. The reader wanting only to obtain a knowledge of the contents of human geography since 1945 should begin at Chapter 2: the academic, too, may want to skip the first section of Chapter 1.

Acknowledgements

The author and publishers wish to thank the following for permission to use or modify copyright material: The Association of American Geographers for figs 3.2 and 3.3; C. W. K. Gleerup Publishers for fig. 5.2; Hutchinson & Co. Ltd for fig. 6.6; Methuen & Co. Ltd for fig. 3.1; Prentice-Hall International for figs 4.2 and 4.3 and *The Statistician* for fig. 3.4.

Preface

Most students reading for a degree or similar qualification are required to pay some attention to their chosen discipline's academic history. The history presented to them commonly ends some time before the present. This has advantages for the historian, because the past is often better interpreted from the detachment of a little distance and there is less chance of hurting scholars still alive. But there is a major disadvantage for the students. In virtually every other component of their courses they will be dealing with the discipline's current literature, and so if their history ends some decades ago, they are presented with the contemporary substance, but not with the contemporary framework, except where this is very clearly derivative of the historical context.

This state of affairs is unfortunate. Students need a conspectus of the current practice in their chosen discipline and should encounter a relevant overview which describes, and perhaps explains too, what scholars believe the philosophy and methodology of the discipline presently are and should be. Such an overview will allow the substantive courses comprising the rest of the degree to be placed in context and appreciated as examples of the disciplinary belief system as well as ends in themselves.

As a discipline expands, so the need for a course in its 'contemporary history' will grow too. In the last few decades, for example, most disciplines, and certainly human geography, have expanded greatly, with expansion measured by the number of active participants and their volume of published work. And the more active members there are, almost certainly the greater the variety of work undertaken, making it difficult for individual students to provide their own conspectus of the discipline from their own reading. Hence the need for a 'contemporary history' course at the present time.

The present book is the outcome of teaching such a course for several years, and is offered as a guide for others, both teachers and students. As with all texts, it has many idiosyncrasies. The course on which it is based is taught to final-year students reading for honours degrees in geography, and is probably best used by people at that level since it assumes familiarity with the concepts and language of human geography. Further, my view is that students probably benefit most from a framework after experiencing some of its contents; this book provides the matrix for organizing the individual parts rather than a series of slots into which parts can be placed later, which would be the case if the course were taught early in the degree.

A series of constraints circumscribes the contents and approach of this contemporary history. First, it deals only with human geography, for several reasons. The most important is that I find the links between physical and human geography tenuous, as those disciplines are currently practised. The major link between them is a sharing of techniques and research procedures, but these are shared with other disciplines too, and are insufficient foundation for a unified discipline. (What price a department of factorial ecology?) Further, my own competence, work and interests lie wholly within human geography and although I have been trained with and by and have worked among physical geographers, and have obtained stimuli from this, I am incompetent to write about physical geography. And finally, much of the human geography discussed here is of North American origin, and many geographers there, especially in the United States, encounter no physical geography as it is understood in British universities. To a considerable extent, therefore, human and physical are separate, if not independent, disciplines. Throughout the book, I use the terms 'geography' and 'human geography' interchangeably.

The second constraint is cultural; the subject matter of the book is Anglo-American human geography. Most of the work discussed emanates from either the United States or the United Kingdom: there are some contributions from workers in Australia, Canada, and New Zealand, but the efforts of geographers in the rest of the world are largely ignored. (A partial exception is Sweden, which has major academic links with the Anglo-American tradition; much Swedish geographical work is published in English.) Such academic parochialism in part reflects personal linguistic deficiencies, but it is not entirely an idiosyncratic decision. Contacts between Anglo-American human geography on the one hand, and, say, that of France and of Germany on the other have been few in recent decades, so to concentrate on the former is not to commit a major error in separating a part from an integrated whole.

The final constraint is temporal, for the book is concerned with Anglo-American human geography during the decades since the Second World War only. Again, this is in part a reflection of personal competence, for I have been personally involved in academic geography for the last twenty years. But the Second World War was a major watershed in so many aspects of history, not least academic practice, and much of the methodology and philosophy currently taught in human geography has been initiated since then.

This book, then, is a history of Anglo-American human geography since 1945. It does not purport to be an objective history, for such an enterprise is impossible. The material included represents subjective judgements: the stress on certain topics is subjective, too, and so is the organization. But although not objective, nor intended to be, the book is neutral. My own opinions are not stated, and are not intendedly implied in anything that has been written (though some of them may be identifiable). There is no commentary, only a presentation of what I perceive to be the salient features.

Arising out of this intended neutrality is a second characteristic of the book, its dependence on the written statements of others. There are few lengthy quotations, but many short ones. Most arguments have, of necessity, been précised or paraphrased, in which case there may be unintentional distortion of the emphasis in the original. My aim has been to report what others have written, and the contents of Chapters 2–6 depend entirely on the published record, for no use has been made of personal memoirs (however valuable these might be, if a representative set could be collected). A consequence of this orientation is a large bibliography. Some authors are commended by reviewers for the utility of their bibliographies. Mine merely indicates the material on which the book is based.

Within the terms of reference set, I have not attempted simply to provide a chronology of arguments. As well as describing the changing contents of Anglo-American human geography I have tried to account for them, to suggest why certain things were said and done, by a particular person in a particular place. Such an attempted 'explanation' of events cannot be neutral, and so it has been handled by writing a book within a book. The outer volume is Chapters 1 and 7; the inner is 2–6: the former presents the subjective account, phrased in terms of the models developed by historians of science, and the latter contains the neutral description. The inner book can be read independently of the outer; the contents of Chapter 7, however, are derived from all of those preceding it.

The account provided in the outer book is not idealist (using that term in the same way as in Chapter 5), for in attempting to explain changes in attitudes to philosophical and methodological topics I rely on my own modelling of scientific progress rather than on the views of the change-agents themselves. The modelling is set in the context of other studies of the history of academic disciplines, notably the physical sciences. By the end of Chapter 7, this model turns out to be something of a straw man. I am not the first to use it within human geography, however, and so my presentation, criticism and then replacement of it represent a contribution to the general enterprise of writing the history of geography (and perhaps of other social sciences).

One problem with writing 'contemporary history' is knowing when to stop. There is a constant flow of new material which can be employed in the task, and there is always the probability (sometimes the certainty, using publishers' advertising) of something very useful appearing tomorrow. Eventually, a halt must be called. For this book, it was in mid-1978, although some later material is referenced (because I was privileged to read it before it went to press). By the time it appears, therefore, this book will of necessity be a period piece, although the arguments in Chapter 7 suggest that a few years will elapse before a major revision is needed.

It was Malcolm Lewis who designed the course of which the material in this book became a part; Stan Gregory was responsible for my teaching it. Both are in no way responsible for the outcome. I am grateful to them for the opportunity to undertake what turned out to be, for me, an enjoyable and stimulating task. The comments (direct and indirect) of the various student audiences have helped to redefine the contents in various ways over the years, and I am grateful for these.

Preparation of this book has been helped materially by several people. My wife Rita has, as always, been of great assistance in a variety of ways, not least in reading the whole manuscript twice and giving much advice on the presentation. Secondly, Walter Freeman has read the entire manuscript and given much of his time in discussing both its contents and its context; I am deeply grateful to him for his interest, his kindness, and his continued friendship. Alan Hay, too, read the complete first draft and commented freely and very helpfully on many aspects of the work. Both he and Walter Freeman felt that there is not enough of me and my opinions in the book: I hope that they understand why. The diagrams were produced by Stephen Frampton and Sheila Ottewell, and Joan Dunn has yet again created an excellent typescript out of a messy manuscript. My thanks to all.

I make it clear in Chapter 1 that the progress of any individual's academic career depends considerably on the actions (and sometimes inactions) of others. No academic is an island. Many people have helped me in the last twenty years, but I should like to express special thanks to Percy Crowe, Walter Freeman, Basil Johnson, Murray Wilson, Barry Johnston, Michael Wise, Stan Gregory and Ron Waters.

R. J. Johnston

1

The nature of an academic discipline

This book is a study of an academic discipline, in particular of its content. But content cannot be discussed fully without an understanding of context, and it is to provide that context that the present chapter has been prepared. In essence, the context is the population involved in the discipline. To study an academic discipline is to study a society, which has a stratification system, a set of rewards and sanctions, and a series of bureaucracies, not to mention a large number of interpersonal conflicts (some academic, some not). To the outsider, academic work may seem to be objective and 'scientific'. But many subjective decisions must be taken: what to study; whether to publish the results; where to publish them and in what form; what to teach; whether to question the work of others publicly; and so on. As with all human decisions, they are made within the constraints set by the society of which the academic worker is a member.

In studying an academic discipline one in effect has to study a society within a society; both are relevant in setting the constraints which are the issue here. The study of the society within the society involves a focus on two questions; 'how is academic life organized?'; and 'how does academic work, basically its research, proceed?'. The first two sections of this chapter are concerned with those two questions. Use of the term 'society within a society' strongly implies that academic life does not proceed independently. It is not a closed system but rather is open to the influences and commands of the wider society which encompasses it. A third question to be asked, therefore, is 'what is the nature of the society which provides the environment for the academic discipline being studied, and how do the two interact?'. This is the subject of the final section of the chapter.

Academic life: the occupational structure

Pursuit of an academic discipline in modern society is part of a career, undertaken for financial and other gains; to most of its practitioners, the career is a professional one, complete with entry rules and behavioural norms. In the initial development of almost all of the academic disciplines studied today, some of the innovators were amateurs, perhaps financing their activities from individual wealth. There are virtually no such amateurs now; very few of the academic publications in human geography are by other than either a professional academic with a training in the discipline or a member of a related discipline. The profession is not geography, however. Rather geography is the discipline professed by the individual academic who probably states his profession as university teacher or some similar term. Indeed, the great majority of academic geographers are teachers in universities or comparable institutions of higher education. (Recent years have seen the invention of new types of higher education institution. The term university is used here as a generic term, therefore. In effect, universities remain the main locations of academic research.) Academic geographers are distinguished from other geographers (many of whom are also teachers) by their commitment to all three of the basic canons of a university; to propagate, protect, and advance knowledge. It is the advancement of knowledge—through the conduct of fundamental research and the publication of its original findings—which identifies an academic discipline; the nature of its teaching follows from the nature of its research.

The academic career structure

Members of university staffs—hereafter termed academics—follow their chosen occupation within a well defined career structure. Two variants of this structure are relevant here: the *British model* and the *American model*. Entry to both is by the same route. The individual must have been a successful student, as an undergraduate and, more especially, as a postgraduate. While the latter, he will have undertaken original research, guided by one or more supervisors who are expert in the relevant specialized field. The results of this research are almost invariably presented as a thesis for a research degree (usually the PhD), which is examined by experts in the field of study. The PhD is an almost obligatory entrance ticket for the American model of the career structure; in the British model it is usual, but far from universal.

While undertaking the research degree, most postgraduate students will have obtained some experience in teaching undergraduates, most particularly in practical and tutorial classes: indeed, some universities, notably those operating the American model, finance their research students through their teaching activities. Having completed the research degree, the individual may then proceed to further research experience, or he may gain appointment to a limited-tenure teaching post in a university department. This appointment may be for a limited period only, offering an 'apprenticeship' in the teaching aspects of the profession whilst either the research degree is completed or the individual's research expertise is consolidated.

Beyond these limited-tenure positions lie the permanent teaching posts, and it is here that the British and American models deviate (Figure

1.1). In the British model, the first level of the career structure is the lectureship. For the first years of that appointment (usually three) the lecturer is on probation: annual reports are made on his progress as teacher, researcher and administrator, and advice is offered on adaptation to the demands of the profession. At the end of the probationary years, the appointment is either terminated or confirmed.

Appointment in this model is almost always to the staff of a department; most departments are named for the discipline which the individual staff member pursues. The prescribed duties involve the undertaking of research and such teaching and administration as directed by the head of the department. The lecturers are on a salary scale, and receive an annual increment; accelerated promotion is possible in some universities. In the United Kingdom there is an 'efficiency bar' after a certain number of years' service; 'crossing the bar' involves a promotion, and is determined by an assessment of the lecturer's research, teaching and administrative activities.

Beyond the lectureship are further grades into which the individual can be promoted. The first of these in the British model is the senior lectureship for which there is no allocation to departments; entry to it is based again on an assessment of the lecturer's conduct in the three main areas of academic life (research, teaching and administration), and is based on open competition across the whole university. The next level is the readership, a position which is generally reserved for scholars of high quality; the criteria for promotion to it focus strongly on research activity by the candidate.

The final grade (still associated with academic departments as against administration of the university as a whole) is the professorship. Although superior in status to the others, this is not simply a promotional category: professorships are appointed to from open competition preceded by public advertisement. Initially the posts of professor and of head of department were synonymous, so a professor was appointed as an administrative head, to provide academic leadership. With the growth in department size in recent years, however, and the development of specialized sub-fields within disciplines, each requiring its own leadership, it has become common for departments to have several professors. In some, the position of head remains with a single appointee; in others, probably the majority, it rotates among the professors; in a few, the headship is now an elected position, independent of the professoriat, to which other grades may aspire.

Finally in the British model, promotion is possible to non-advertised, personal professorships. These positions are relatively rare, and are reserved for distinguished scholars who have risen through the ranks of the one establishment and for some reason have not obtained appointment to a professorship (a 'chair') in the normal manner.

The system under the American model is similar, but generally simpler (Figure 1.1). There are three permanent staff categories: assistant professor, associate professor, and professor. The first contains two sub-categories—those staff on probation who do not have 'tenure' and those with security of tenure. Each category has its own salary scale but no automatic annual increments. (In the United States, these scales vary from university to university—according to prestige, state etc.) Salary adjustments are the result of personal bargaining, on the basis of academic activity in the three areas already listed, and as a result it is possible for the salary scales of the categories to overlap in the same department. Movement from one category to another is promotional, and is a recognition of academic excellence. Professorships are simply the highest promotional grade and do not carry any obligatory major administrative tasks. Heads of departments in the American model are separately appointed, usually for a limited period only, and they need not be professors (usually referred to as 'full professors').

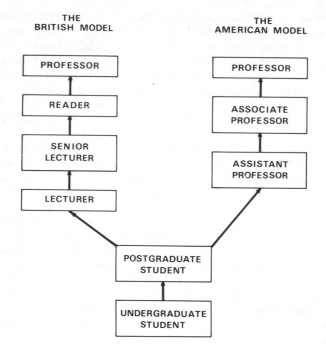

Figure 1.1   The academic career ladder. Note that whereas in the American model it is very unusual not to progress up through the stages in an orderly sequence, in the British model missing steps is quite common (i.e. some lecturers move direct to readerships, even to professorships, without first being senior lecturers—there are no senior lecturers in the 'Oxbridge' universities—and many professors are recruited from the senior lecturer rather than the reader grade)

Status, rewards, and promotion prospects
The rewards of the academic profession are the salary levels, plus the general status of the profession, the particular status of its individual promotional categories, and the relative independence and flexibility of the working conditions. The tangible rewards—the payments—are determined by the academic's category and his position within it. Promotion must be 'earned'. How?

Academic work has the three main components—research, teaching and administration. Entrants to the profession have little or no experience of academic administration and their only teaching has probably been as assistants to others. Thus it is very largely on proven research ability that the potential of the aspirant for an academic career is judged. To some degree, research ability can be equated with probable teaching and administrative competence, since all require the same personal qualities—enthusiasm, orderliness, incisive thinking, and ability to communicate orally and in writing; to a considerable extent, however, academics are appointed to their first position on faith, hence the usual probationary periods before tenure is granted.

Once admitted to the profession, the academic will have to undertake all three types of work, so that promotional prospects can be more widely assessed. In effect, teaching and administration probably carry less weight with those responsible for promotions than does research. This is partly because of difficulties in assessing performance in the first two, and partly because of a general academic ethos which gives prime place to research activity in the evaluation of peers.

Although not necessarily progressive in terms of increasing level of difficulty, administrative tasks tend to be more complex and demanding of political and personal judgement and skills as the academic becomes more senior. As made clear by the 'Peter Principle' (Peter and Hull, 1969), a person's ability to undertake a certain task can often only be fully assessed after he has been promoted to the relevant position. Promotion must frequently be based on perceived potential, and although it is possible to point to somebody whose administrative skills are insubstantial it is not always easy to assess who will be able to cope with the more demanding administrative tasks.

It has long been realized that assessment of teaching ability is difficult (although clear inability is often very apparent). Student and peer evaluations are possible and useful, as is scrutiny of the work done by the students taught. But criteria for judging teaching performance at university level are ill-defined, and what is expected from a lecturer/tutor/ seminar leader often varies quite considerably within even a small group of students. Thus the majority of teachers are usually accepted as competent and undistinguished, and their performance is neither help nor hindrance to their promotion prospects.

And so the main criterion for promotion is usually research

performance, although a relatively undistinguished record in this area can be compensated by excellence elsewhere. How then is research ability judged? Details of how research is undertaken are considered in the next section; here the concern is with its assessment rather than with stimulus and substance.

Successful research involves making an original contribution to a field of knowledge, in a variety of ways. It may involve the collection, presentation, and analysis of new facts, within an accepted framework; it may be the development of new ways of collecting, analysing and presenting facts; it may comprise the development of a new way of ordering facts—a new theory or hypothesis; or it may be some combination of all three. The originality of a contribution is judged through its acceptance, or validation, by those of proven expertise in the particular field. The generally accepted validation procedure is publication, hence adages such as 'unpublished research isn't research', 'publish and be damned', and 'publish or perish'.

The outlets for the publication of research findings are the scholarly journals, which operate fairly standard procedures for the scrutiny of submitted contributions. (There is general belief that the journals published by academic societies tend to have higher standards than those published by commercial companies, but the validity of this is difficult to assess.) Manuscripts are submitted to the editor, who seeks the advice of qualified, often senior, academics on the merits of the contribution; these referees will recommend either publication, rejection, or revision and resubmission. When accepted, a manuscript will enter the publication queue, which in some cases may be up to two years long, depending on the academic reputation of the journal.

Although widely accepted, this procedure has certain inbuilt flaws, largely because it is operated by human decision-makers. The opinions of both editor (on the manuscript, and on whom he chooses as referees) and referees may be biased in some way, so that a paper can be rejected by one journal but accepted by another, without alteration. In most disciplines there is an informal ranking of journals in terms of their prestige, and it is considered more desirable to publish in some rather than others. (This prestige ranking is sometimes taken into consideration by appointment and promotion committees.)

Some research results are published in book form, rather than as articles in journals. Most academic books are texts, however, published by commercial companies whose main interest is marketability among the large student population. The textbook may be innovative in the way that it orders and presents material, and can be beneficial to its author's reputation (as well as to his bank balance), but it is not usually a vehicle for demonstrating research ability. Many companies, including the university presses, do publish research monographs, however, to present the results of major research projects to relatively small academic markets. Their decisions on whether to publish are made on academic as well as commercial grounds, usually with the aid of academic referees, and their output is validated through the book review columns of the journals.

Processes of promotion and appointment: patronage
Whatever the weighting given by the relevant committees to the three main academic activities (there are others pursued by some, such as consulting for outside bodies), there is still a major question of how such committees make their assessments. The only 'objective' information which can be presented to them are examination (degree) results and lists of publications which have been validated by academic journals. But how is such information to be evaluated?

Two modes of assessment are widely used and relied upon: the written opinion of a third party (a referee) and the interview. In the British model, considerable weight is placed on the former. An applicant for a position must supply a list of referees who will provide an opinion on his suitability for the post; fairly clearly, he is likely to ask those who he has worked with, and who are favourably inclined towards him, to act in this capacity. As appointment committees tend to be swayed very much by these reports, especially in their preparation of a shortlist of candidates to be interviewed, the opinion of well respected members of the discipline who act as referees are often crucial. Thus certain leaders in a subject often find it easier to get their candidates appointed to posts than do others: there is a considerable element of patronage involved in obtaining a university post, especially a first university post. This patronage is particularly important in getting the candidate onto the shortlist of those given an interview, from which the final selection is made.

Promotions under the British model are largely based on the opinion of one person, the head of the candidate's department. Reports must be made on each lecturer: annually during the probation period, at the confirmation stage after probation, on reaching the 'efficiency bar', and for either accelerated promotion within the lecturer scale or promotion to senior lecturer. Again, therefore, there is a strong element of patronage, although constraints are built into the committee system to try and ensure fairness for all, including the right either to present one's own case or to appeal against a decision. Some universities use outside referees to provide extra evidence for cases of promotion to senior lecturer, and promotions to readerships always involve the use of external referees, usually nominated by the head of department concerned.

Referees and interviews are also used in the appointment of professors. Two sets of referees are used. The first is those nominated by the candidate as being prepared to provide a confidential report on his potential for the job in question; the second is a group of assessors nominated by the university, usually senior academics in the relevant field. The latter not only comment on the list of candidates who have applied but also suggest other worthy candidates who might be approached: their potential patronage power is great.

Procedures are slightly different under the American model. More weight is usually given to an enlarged interview, often involving candidates in giving seminars to the department and meeting with various groups of staff. The reference is still important, however, especially for aspirants to first teaching positions, and a letter from a respected leader in a field can be very influential in gaining appointment for his former student. In general, the individual candidate is more active himself in this system, however, perhaps canvassing for interviews during the annual conferences of scholarly societies, for example. For promotions and for salary rises there is considerable bargaining between individual, department chairman, and university administrators: external referees' opinions may be sought when tenure is being confirmed, but not usually thereafter.

In all of these procedures the applicant, for a new position or for a promotion, depends to a considerable extent on the opinions of senior academic colleagues in the evaluation of his prospects. Some opinions carry more weight than others, and so it is important when developing a career to identify potentially influential individuals, to keep them informed of your work, and to enlist their support for your advancement. Because of the lack of truly objective criteria for measuring research, teaching and administrative ability, such patronage is crucial.

Other rewards and the sources of status

The tangible rewards of an academic career are the salary and the life style, plus the occasional extra earnings that are possible—for examining, writing, lecturing and consulting. In addition the hours are flexible; the possibilities for travel are considerable; and the constraints on when, where and how work is done are relatively few, compared to most other professions. And there are other, less tangible rewards. Involvement with the intellectual development of students is a considerable reward for many, and brings much satisfaction. There is also the charisma associated with recognized excellent teachers, and even more so by leading researchers, whose publications are widely read, whose invitations to give outside lectures are many, and whose opinions as examiners, referees and reviewers are widely canvassed. And the conduct of research brings its own rewards, apart from the charisma; the satisfaction of having identified and solved a significant problem is often considerable, and a major reward for the dedicated researcher.

One reward common to social systems exists in the academic community too: power. Patronage is power, as is any work as examiner, referee, or reviewer. The careers of others are being affected, and exercise of this power can bring with it the loyalty and respect of those who benefit.

Because the academic system is so dependent on the opinions of individuals, because the opinions of some individuals are more valued than are those of others, and because power over others is a 'commodity' widely desired in most societies, many academics seek positions of influence.

Among the most influential positions within an academic discipline are the administrative headships of departments. Holders of these positions can instruct other staff members (usually after consultation) in their teaching and administrative duties; they are frequently used by staff and students as referees for job and other (such as research grant) applications; and their reports are crucial in the promotions procedure. The departmental organization of universities is a bureaucratic device which makes for relative ease in administering what is often a very large institution. (It also tends to fossilize disciplinary boundaries, as will be discussed later.) Departmental heads not only have power over members of their own discipline's staff, they also participate in the administration of the university as a whole, and they treat within the university-committee system for departmental resources. As in all bureaucracies, there is a tendency for the status and power of the various departmental heads to be a function of the size of their 'empires' (Tullock, 1976). Heads of large departments, especially large and growing departments (growth being generally considered as a 'good thing'), are very often the most important individuals in a bureaucracy. Thus they have the incentive, whether or not they are permanent occupants of the headship, to build up their departments, which usually means increasing student numbers, since universities tend to allocate resources to departments according to the numbers of students taught. This brings power and prestige, both within the university and beyond; it is an added reward for the academic bureaucrat, and the power over resources which it involves is usually of benefit to his whole department.

Finally, the academic bureaucrat, whether or not a head of department, can gain power beyond his home university through, for example, obtaining positions on committees. These may be concerned with his subject through professional bodies, with the allocation of research moneys by public or private foundations, or with a wide range of public duties. Again, the status and power obtained can spill over to others, since patronage is important in all of these roles.

The academic working environment

The nature of an academic discipline is defined by the phenomena which it studies and the ways in which it investigates them; the contributions of individuals comprise the research that they undertake and their efforts to integrate their findings with the discipline's general body of knowledge. Thus a study of how a discipline proceeds needs to understand how the body of knowledge is structured and how research investigations are organized within this constraint. Such understanding is far from complete, and indeed the study of the operations of academia has become a major research task in itself only relatively recently. The present brief discussion is based on a few statements regarding the structure and operation of scientific disciplines (notably Mulkay, 1975 and Mulkay, Gilbert and Woolgar, 1975). Whether such statements, which are largely based on studies of 'big science', such as physics, are relevant to a social science such as human geography is debatable, and this point will be considered more in the final chapter. For the present, these general formulations offer a useful framework for the study of the nature of a discipline such as human geography: they have indeed been used by others for that task (e.g. Haggett and Chorley, 1967; Harvey, 1973).

The progress of a science
To the casual outside observer the structure, content and methods of working of an academic discipline, particularly of a science, are somewhat mysterious. The aim of the discipline is the advancement of knowledge, which can be achieved by a variety of methods, including experimentation. But how does the academic go about his work?

Outside views of science usually omit any attention to the fact that an academic community is one in which conflict is often rife, and that although the general tenets of academic good conduct may be held by all, there are many interpretations of how research should be done, what is good research, what results are valid, and so on. (They could be somewhat disabused of these general beliefs in the purity of science by reading either a general history of a scientific discovery—e.g. Watson, 1968—or a good novel on the scientific enterprise—e.g. Cooper, 1952.) The values usually ascribed to academia, and in general widely accepted by the academic community, are, according to Mulkay (1975, p. 510):

1   The norm of originality; academics are forever striving to advance knowledge, to discover and explain new aspects of the world in which they live, and to conduct original research in pursuit of this aim;

2   The norm of communality; all information is shared within the community, being transferred through the accepted modes of communication (notably the academic journal), whilst its provenance is always recognized;

3   The norm of disinterestedness; academics are devoted to their subject, and the main reward that they seek is the satisfaction of knowing that they have assisted in advancing knowledge;

4   The norm of universalism; judgements of the work of others are entirely impartial and are based on its academic merits alone; and

5   The norm of organized scepticism; knowledge is furthered by constructive criticism, so academics are always reconsidering both their own work and that of others.

Thus, according to these five norms, academic work is carried out in a neutral fashion: there is a complete lack of partiality, self-seeking, secrecy and intellectual prejudice. Such norms assume the existence of objective criteria, of perfect ability to know all relevant work and to be able to assess it properly, and of humility.

Even if these five norms do provide an accurate description of the operation of the scientific enterprise, they offer no answer to the crucial question 'what do academics decide to work on?'. Most work involves searching for order, identifying laws of how things happen and gaining understanding of particular events (the two are, of course, closely intertwined). But the range of possible topics which could be studied is almost limitless. In the initial stages of development of an academic field, when little previous research has been done, one individual might expect to handle the whole field. But as more and more research is reported, so each individual can expect to master only a small portion of it. Specialization ensues. From a broadly based general education, which then narrows into a more specialized undergraduate education, the postgraduate comes to focus on one particular set of research questions only—usually that espoused by the group of scholars with which he chooses to work. He thus moves away from the general textbooks, which codify the accepted knowledge, into the specialist research literature, and concentrates on a very particular field.

This socialization of academic researchers into specialisms is very much a function of the expansion of all academic disciplines in recent years, and the great growth in the volume of research literature (on geography as a whole, see Stoddart, 1967a). Within their specialism, they learn the valid findings of previous research, how to formulate sensible further research arising out of this literature, and how to set about conducting this research.

Normal science

Each group of scholars working in a particular field becomes a small, often somewhat isolated, community of researchers and teachers, within which individuals communicate and conform. They operate a set of theoretical, methodological and empirical procedures which are widely accepted and their own contributions involve small additions to the community's theories by using its methods to answer new questions; occasionally one or more of them may decide to assemble the known material—both findings and methods—in a textbook. Such communities and their procedures have been termed *paradigms* by a historian of science, Kuhn (1962): 'a paradigm is what the members of a scientific community share, *and*, conversely, a scientific community consists of men who share a paradigm' (Kuhn, 1970, p. 176). What this means, as Popper (1959) expresses it, is that a scientist, once socialized into his research field, can proceed direct to the heart of his problem; there is an existing structure of doctrines within which that problem is set and which provides the framework and methodology for his analyses. He does not begin his research *de novo*, but on the basis of a large volume of accepted research.

Identification of these research communities involves a number of procedures aimed at separating out the 'invisible colleges' of practitioners who work together (Crane, 1972). Attendance at specialized conferences; the private circulation of papers in prepublished form; citations in the works of others: all of these give clues as to the existence of communities involved in the same research task. Within them individuals receive recognition of the value of their work, and find the patrons who will assist in their career advancement; some communities are based on only one or a few 'super-patrons'—the paradigm leaders. In order to get the rewards of recognition and patronage, the individual must conform to its paradigmatic norms (Mulkay, 1975, p. 515):

> It is clear that the quality or significance of a scientist's work is judged in relation to the existing set of scientific assumptions and expectations. Thus, whereas radical departures from a well defined intellectual framework are unlikely to be granted recognition early under normal circumstances, original contributions which conform to established preconceptions will be quickly rewarded.

Thus the main norm of academic life would seem to be not one of the five listed earlier but rather conformity. The contribution of each separate piece of research is that it adds a little more to the paradigm's established body of knowledge; the existence of this 'fact' which is being contributed is predictable, according to Kuhn (1962), so that the research merely confirms the validity of the paradigm, laying the groundwork for the next prediction.

The operation of this set of procedures is called *normal science*:

> Perhaps the most striking feature of the normal research problems . . . is how little they aim to produce major novelties, conceptual or phenomenal. Sometimes . . . everything but the most esoteric detail of the result is known in advance . . . the range of anticipated, and thus assimilable, results is always small compared with the range that imagination can conceive . . . the aim of normal science is not major substantive novelties . . . the results gained in normal research are significant because they add to the scope and precision with which the paradigm can be applied. . . . Though its outcome can be anticipated, often in detail so great that what remains to be known is itself uninteresting, the way to achieve that outcome remains very much in doubt. . . . The man who succeeds proves himself an expert puzzle-solver (Kuhn, 1962, pp. 35–6).

Once a paradigm has been defined, therefore, the researcher has available to him:

1   an accepted body of facts, ordered and interpreted in a particular way;

2   an indication of the puzzles which remain to be solved; and
3   a set of procedures with which to approach puzzle solving. In more
general language, the concept of normal science places each individual
researcher within a fairly deep rut and sets him to solve some relatively
trivial problems, from which the new facts will add very slightly to the
store of knowledge and the accuracy of the predictions from the paradigm's theories.

Scientists are not perfect beings, however, and they cannot predict for
all eventualities. While the process of normal science continues, slowly
accumulating extra information to build into the store of knowledge,
occasionally it will throw up anomalies, findings which do not fit the
paradigm's expectations. These discovered anomalies must be accounted
for, and their existence built into the theory which they in part
invalidate; thus in most normal science situations, the discovery of
anomalies, often only very small, will lead to paradigm adaptation.
Occasionally, however, an anomaly will be produced that cannot be
incorporated into the paradigm, because it contradicts so much of what
has been previously accepted as fact. Initial reactions often suggest that
such anomalies should be ignored, in the hope that they are chance
results of a misconducted experiment. But if they persist, then the
problem that they pose must be recognized, and the paradigm is thrown
into a state of crisis. And so people begin to work on the problem, often
while others continue the general activity of the paradigm under question, and attempts are made to forge new theories which will be the bases
for alternative paradigms that incorporate the anomaly and reinterpret
the earlier accepted facts. This Kuhn (1962, pp. 89–90) terms 'extraordinary research', and notes

> Almost always the men who achieve these fundamental inventions of a new
> paradigm have been either very young or very new to the field whose para-
> digm they change . . . obviously these are the men who, being little commit-
> ted by prior practice to the traditional rules of normal science, are particularly
> likely to see that those rules no longer define a playable game and to conceive
> another set that can replace them.

Once what seems to be an acceptable new paradigm has been produced,
then the relevant scientific community must decide between the two
(Kuhn notes—p. 110—that since no paradigm is perfect and offers a
solution to all problems there is always a judgement to be made). If the
new paradigm is accepted, then a revolution in scientific thought has
taken place, and a new view of the world, or some component of it, has
been produced. Often, again as Kuhn (p. 139) notes, scientists tend to
write the history of their subjects backwards, so that the revolutions are
made to seem like linear developments cumulating to the present situation; this also has the benefit of reducing dissensus between the protagonists of alternative world views.

According to this model of normal science, therefore, the practitioners in a discipline adopt a particular method of looking at their subject matter and of solving problems involved in explanation of its content. Their problem-solving proceeds in a steady, cumulative manner, with occasional modifications to their general view in order to accommodate anomalies which they discover. Very occasionally, however, they encounter a large anomaly which cannot be incorporated into their world view. Some practitioners then have to devote their energies to writing a new paradigm which will embrace this anomaly, and which of necessity has to reinterpret the old material. When this is achieved, and the achievement is recognized, a revolution takes place in the nature of the discipline. In a sentence, therefore, science proceeds in a steady fashion, along well trodden paths, with occasional major breaks in the continuity marked by important changes in organization of material, definition of new problems, and techniques for problem-solving.

Criticisms of normal science

Kuhn's model is itself an attempt to provide a paradigm, and its application throws up anomalies and invites criticism, both constructive and destructive. Not surprisingly, there is a considerable literature of such criticisms, some of which have been generated by Kuhn's own presentation of the model. Thus Masterman (1970), for example, points out that Kuhn uses the term paradigm in no less than twenty-one different ways, although fortunately these can be amalgamated into three major types of usage:
1   the metaphysical, representing a 'world view';
2   the sociological, which provides a set of scientific habits; and
3   the construct, which is a classic work that provides the tools and procedures for later investigations. Together, they indicate that normal science involves working not only within a prescribed scientific framework but also in a given social structure.

Among the major questions that remain unanswered with regard to Kuhn's model is 'how are the paradigms initially defined?'. Before communities based on particular paradigms are formed, disciplines occupy a pre-paradigmatic state (sometimes termed proto-science) in which research is not cumulative and practitioners are seeking to form a framework for their research: facts are forever being reinterpreted and there is no constant body of knowledge. Eventually, one attempt at a paradigm seems to offer the best chance of progress and becomes widely adopted; very often such a paradigm is founded by a charismatic personality whose leadership results in particular questions being asked and procedures employed. Thus the practice of science can be dominated by personalities and criticisms are often taken as personal attacks.

A second, and probably more important, question concerns the nature of revolutions: 'how are paradigm changes brought about?' (see, for example, Harvey, 1973, pp. 120–8). Many serious anomalies may not be

recognized for what are, leading Watkins (1970, p. 33) to call science the
'scientist's religion'; the scientist is blinkered so that he will only accept
the faith, like a theologian who cannot reconcile two biblical passages.
But some individuals, at some times, will recognize the importance of the
anomalies, and attempt to provide new paradigms. What determines
when and how this takes place? Is it again the consequence of a com-
manding personality, or, as discussed in the next section, does it result
from demands generated by the external environment?

Watkins (1970) has looked in detail at the proposed revolutionary
process, and suggested flaws in Kuhn's arguments. He identified five
component theses to the model:

1 The paradigm-monopoly thesis, which states that all scientists work
at any one time under a single paradigm only;
2 The no-interregnum thesis, which states that scientists make almost
instantaneous changes, with no major floundering period while they
decide on the relative merits of competing paradigms;
3 The incompatibility thesis, which states that no paradigm is compat-
ible with the one it supersedes;
4 The gestalt-switch paradigm, which states that the switch from one to
another in a scientist's activities is immediate; and
5 The instant-paradigm thesis, which follows from the previous ones
and implies that new paradigms must be invented relatively rapidly.

Since, he argues, the last is demonstrably untrue, then most academic
disciplines are likely to be in a constant state of at least dual-paradigm, if
not multi-paradigm debate.

If Watkins is correct, this suggests the need to develop a model of
academic disciplines which incorporates almost continual inter-
paradigm debate. Such a development will probably need to incorporate
the sort of typology of researchers developed by Eilon (1975) to characterize management scientists. His categories are:

1 The chronicler, whose role is the description of reality as he has been
trained to see it within the confines of the paradigm to which he sub-
scribes;
2 The dialectician, who interacts with the system he is describing: he
'believes it is necessary to debate and argue issues in order to elicit the
facts . . . challenging stated views or records, in order to uncover what
otherwise may remain hidden from an innocent observer' (Eilon, 1975,
p. 361);
3 The puzzle-solver, who seeks a solution to problems defined by the
established paradigm and who does not question the validity of seeking a
solution to that puzzle;
4 The empiricist, who, like the chronicler, seeks to describe reality,
using a particular set of tools;
5 The classifier, who takes information collected by others and orders it
within the framework of the accepted paradigm;

6   The iconoclast, who destroys cherished beliefs, the assumptions on which they are built, the deductions from certain assumptions, the conclusions and interpretations, and the incompatibilities between theory and practice, between the predicted and observed worlds; some iconoclasts may be destructive only, whereas others are agents of paradigmatic revolutions; and

7   The change-agent, who is concerned not with the establishment of new facts and theories but rather with the use of existing knowledge to change the system under consideration and thereby produce a 'better world' on the basis of academic understanding.

Individual workers are not necessarily always centrally sited in one of these archetypes, and many will work in one or more areas, perhaps even contemporaneously. Probably the majority of academics in any discipline are in one or more of the types which are characteristic of normal science, however—the chroniclers, the puzzle-solvers, the empiricists, and the classifiers (Mercer, 1977). The change-agents, too, are applied normal scientists: most, if not all, of their published work will be supportive of the dominant paradigm at the time when they trained as scientists and their status within their profession reflects both the quality and the quantity of their published work. Iconoclasts, and especially constructive iconoclasts, are few in number, and relatively few academics aspire to the status of founders of a new paradigm.

Political revolutions occasionally take long periods of time to achieve, as an increasing proportion of the population is won over to the new view; most occur very rapidly, however, with opposition being subjugated, even removed. The latter course is not feasible within the academic community and change must be achieved by persuasion. Even the most gifted iconoclasts usually face considerable resistance, because many scholars have their careers invested in a certain paradigm and could lose status and power (actual and potential) if a revolution in thought took place. Thus the battle to launch a new paradigm often occurs among the younger generation of scholars, who can more readily be made aware of the anomalies produced by the currently accepted paradigm and can see possible benefits from associating themselves with potentially powerful patrons.

As a final comment on Kuhn's model, it should be noted here that, while accepting the existence of what Kuhn calls normal science, Popper believes that true science is what Kuhn terms extraordinary research (see above, p. 17). To him, the normal scientist is really an applied scientist:

> 'Normal' science . . . is the activity of the non-revolutionary, or more precisely, the not-too-critical professional: of the science student who accepts the ruling dogma of the day; who does not wish to challenge it; and who accepts a new revolutionary theory only if almost everybody else is ready to accept it—if it becomes fashionable by a kind of bandwagon effect. To resist a new fashion needs perhaps as much courage as was needed to bring in about. . . . The

success of the 'normal' scientist consists, entirely, in showing that the ruling theory can be properly and satisfactorily applied in order to reach a solution of the puzzle in question (Popper, 1970, pp. 52–3).

To Popper, it is the 'extraordinary' researcher who is the real scientist, for with such people 'there was, ever since antiquity, constant and fruitful discussion between the competing dominant theories of matter . . . the method of science is, normally, that of bold conjectures and criticism' (Popper, 1970, p. 55).

Any history of a discipline, therefore, must recognize this dichotomy between normal and extraordinary research. It will accept that when paradigms are established they will be inhabited by many normal scientists who will add to the store of knowledge in uncontroversial ways. But the encounters with anomalies will ensure the existence of iconoclasts who question the existing paradigms, propose alternatives, and sometimes succeed in establishing a revolution in scientific practice. Although the normal scientists may form the majority of the adherents of a discipline, it is the works of the minority of iconoclasts on which a history such as the present book will focus.

Branches of paradigms
Rapprochement between competing paradigms is unlikely to be rapid in most situations, so that during a period of revolution, or of multiple-paradigm science, a discipline is likely to split into separate sub-communities, focused on particular views, and perhaps leaders. These communities are identifiable as invisible colleges (see p. 16). If strong enough, and if they encounter sufficient opposition from the establishment which dominates with other paradigms (the major figures in such establishments will be the editors of the important journals who act as the gatekeepers to the discipline's main scholarly outlets), they may launch their own journals and monograph series. This could lead eventually to a permanent split and the establishment of a new discipline, and there are several examples of this. But in most cases such fission is unlikely. Separate departments competing for the same students, to approach certain subject-matter in different ways, are likely to create confusion, and perhaps failure for both. More importantly, such a schism would be against the interests of the disciplinary bureaucracy, which is in the hands of the established paradigm. A new department forged out of the old would lead to a decline in the latter's size, and thus the status and power of its head. To avoid this, the latter is likely to compromise his championing of one paradigm and instead show 'liberality' by allowing the opposition to set up a branch within the established order. By co-opting the new group in this way, the power of the bureaucracy is maintained, even enhanced. Eventually, the members of the offshoot branch may rise to positions of bureaucratic power and status within the career structure so that their paradigm, through a process of generational

succession, becomes the dominant while the old fades into a minor component.

According to this view, potential revolutions result in a series of branches within a discipline rather than either fission or the replacement of one paradigm by another (the latter usually occurs, but slowly as a feature of population-ageing rather than as a consequence of scientific debate). Mulkay (1975, p. 520) has presented another case for the development of intra-discipline branches:

> In science, new problem areas are regularly created and associated social networks formed. . . . The onset of growth in a new area typically follows the perception, by scientists already at work in one or more existing areas, of unresolved problems, unexpected observations or unusual technical advances, the pursuit of which lies outside their present field. Thus the exploitation of a new area is usually set in motion by a process of scientific migration.

Within an accepted paradigm, therefore, new areas of study are initiated, and some students and researchers move into them, perhaps only for a few years while the main problems are sorted out. Such branches may be identifiable by their particular methods and techniques, others by their subject matter. These branches can probably coexist relatively easily and share in some activities, such as basic training for students. Academics will migrate into new areas of study with some ease, without in any way compromising their basic scientific stance, and perhaps move out just as easily as they see that a particular academic seam is being worked out. Within Eilon's typology these are probably a sub-category of the puzzle-solvers, comprising competent researchers able to assimilate new material fairly readily, under active leadership, but unable to make the breaks which characterize iconoclasts.

The argument for the branching model of disciplinary development is suported by Toulmin (1970, p. 45) who notes that, with the growing number of academics, incongruities between theory and reality are frequently recognized, so that

> perhaps every new generation of scientists having any original ideas or 'slant' of its own finds itself, at certain points and in certain respects, at cross-purposes with the immediately preceding generation.

If those differences are large, a new paradigm may emerge; if they are small, an extra branch of the discipline's trunk is the likely consequence. Because most disciplines are now several decades old, have a considerable history of research activity, and comprise many active researchers, most academic departments will comprise a strong basic trunk of accepted methodologies, a number of separate branches representing the specialisms within the dominant paradigm, and perhaps a few dissident branches, fomenting potential academic revolution.

The external environment

So far in this discussion, academic disciplines have been presented as autonomous units, which make their own decisions about what to study, and how, without any outside influence. Clearly this is an unreal position. Academic research is no longer conducted in monasteries by individuals who are not inhabitants of the 'world outside'. Most work in universities, and their duties include the instruction of students; customer-demands must be recognized, if not entirely met, and the content of teaching is likely to interact with the nature of research. And most universities are paid for, directly in grants and indirectly through subsidized student fees, by governments who represent the wider society, and these will want to influence, if not to determine, the content of academic teaching and research: a few universities are privately financed, and they must convince donors, as well as students, of the relevance of their work to current societal concerns. Thus to study the history of an academic discipline requires attention to the framework within which it operates; an outline of the framework for human geography in the decades since World War II is presented in this final section of the introduction.

World War II is more than a convenient period from which to commence this history of human geography; it marked a major watershed in the development of the societies which are the focus of the book—the United Kingdom and the United States. It cannot be considered in isolation, however; just as important for the present discussion are the world-wide economic depression which preceded it and the Cold War which followed.

For the first time a major international conflict was not determined solely by the sacrifices of men in battles on land and sea, although there were many such sacrifices during World War II. And the extra dimensions of this war did not just involve the development of air space as a further arena for conflict. The war was fought not only between military forces with guns, but also between scientists, and victory was hastened, if not ensured, by the scientific superiority of the Allied Powers, most obviously at Hiroshima and Nagasaki. It marked the 'coming-of-age' of science and technology. For long these had been major elements in the developing industrialization of the western world, but their dominance was established during the years of conflict, and there was to be no retreat from the many technological advances made by the researchers who assisted in the military effort. Thus the war heralded the era of science and technology, of the dominance of the machine.

Associated with this growth in activity, and also in prestige (with governments and with society at large), was a parallel development of what has become known as social engineering. The major economic depression of the 1930s, finally triggered by the Wall Street crash of

1929, had a massive impact on the nature of government activity and led to the initiation of many measures aimed at the relief of poverty and deprivation and the assuaging of the liberal conscience. In the United States, this was represented by the New Deal legislation of President Roosevelt's governments, which aimed at relief and encouragement to industry and, through the Social Security Act, public support for those who, by no fault of their own, were indigent. Similar measures were introduced by the National Government in Britain; more were fore-shadowed by the plans of the war years, such as those promoted in the Beveridge Report on social security, and the landslide victory of the Labour Party in 1945 heralded the introduction of many social democrat policies aimed at giving government a much greater peacetime role in the organization of the economy and society than had ever previously been envisaged.

This development of social engineering was associated with a rising status for the social sciences, and a great expansion in their activities. Economics was the first to achieve prestige, through the contributions of Keynes and others to the solution of the problems of the depression, the organization of the economies during wartime, and participation in the planning of a new world economic order after the war. Others followed. Social psychological research was widely used in the evaluation of per-sonnel by the armed forces and after the war opinion surveys proved valuable to politicians and related groups while market research was increasingly used by industry (along with psychology in its advertising efforts). All of these fields adopted the 'scientific methods' of the more prestigious hard sciences, and their successes were emulated by other disciplines which saw roles for themselves in the 'new Jerusalems', such as sociology, social administration, and, later, geography. To be sci-entific was to be respectable and useful.

The war years saw the end of the economic deprivations of the depression, as manufacturing output was boosted to provide the machin-ery of war. During the cold-war period that followed, military produc-tion remained considerable and kept many people in work. Further, there were many years of doing-without to be compensated for, and with full employment, government direction of, and increasing involvement in, economic affairs, and the need to re-equip industries, the two decades following the war were characterized by an economic boom in the western world. Apart from the greater government involvement, this era in industrial development was marked by another major new characteris-tic, the development of the giant firm, including the multi-national. The concentration and centralization of capital proceeded rapidly, probably more rapidly than during any preceding period; the average size of firms and factories increased and the economy of the world became dominated by a relatively small number of concerns, which were usually encouraged in their activities by governments.

Rebuilding the ravaged war arenas of Europe also placed new demands on societies, and the planning profession emerged from earlier obscurity to take on a major role in preparing the blueprint for a new social order. The need for such action had been realized during the war years in the United Kingdom with the preparation of a series of reports concerned with future land-use patterns, and with the spatial distribution of economic activity, at local and regional scales. Cities were to be rebuilt; New Towns were to be constructed; a more balanced inter-regional distribution of industry and employment was to be ensured; agricultural land was to be protected; and residential environments were to be improved: all of this made great demands on social scientists, as well as engineers. The greater degree of commitment to the private ownership of land in the United States led to a slightly slower acceptance of the need for spatial planning there, but its heyday came with the rapid growth of problems involved in catering for the upsurge of ownership and use of the automobile: transport planning and engineering soon became growth industries, allied with the automobile industry and the companies which constructed the major highways. Again, this became a major sphere of government activity.

For all of these tasks—economic growth and planning; spatial planning; social administration; technological change; management etc.— there was a need for educated personnel, and the universities received unprecedented demands for their graduates to serve the new needs of society. Education expanded rapidly. The existing universities and colleges grew and many more were founded. Science and social-science departments expanded to meet the need for more students. The extra staff were involved in research, which increased the tempo of paradigm-development and questioning, and the products of the research were wanted by governments and businesses to assist in the achievements of their aims. Rather than the places where a small elite were educated and a few privileged individuals followed their research interests, the universities became centres of society's development—the 'white-hot technological revolution' which Harold Wilson promised the British in the early 1960s. Research projects became bigger, supported by large grants from outside bodies and carried out by specially employed graduates, and the rate and volume of publication increased exponentially.

The period from 1945 to about 1965 can be characterized as the years of scientific dominance, therefore, but in the mid-1960s there was the beginning of doubt, of a questioning that the world had been put to rights and that the problems of scarcity and hardship had been solved; the only remaining problem was to ensure that everybody participated in the prosperity. The seeds of doubt were many. They included, notably in the United States although not only there, the problems of war, particularly the increasingly unpopular conflict in Vietnam and neighbouring countries. There was also a recognition that poverty was not being alleviated, neither globally nor within the 'advanced' countries: if anything the disparities between rich and poor were increasing, and, with the extension of western medicine to what became known as the 'Third World', more and more people were being born into a world of permanent deprivation. The particular situation of certain groups, both minority—particularly racial—and majority—women, became a focus, again notably in the United States with the great surge in support for he civil-rights campaign, while a few scientists and associated activists kindled intense interest in the problems of resource exhaustion, environmental conservation, and the potential for ecocatastrophe. Finally, there was a growing concern that the growth of big organizations, both private and public, was leading to increased alienation at work and problems of identity.

All of these issues troubled the liberal conscience throughout the period under review; they were brought to the forefront of public attention in the late 1960s. In this, much of the leadership came from the student body, especially from social-science students, and the late 1960s saw major confrontations between students and other elements of society, in Berkeley, London, Paris, Chicago and elsewhere. Initially much of the protest was couched within the liberal, social-democrat ethos of the western world: its consequences included a questioning of the ethos of economic growth, of big business and big government, and of the contributions of science and technology to human dignity. In the universities, the sciences and associated technological subjects declined rapidly in popularity, and many departments were unable to fill their student quotas. Meanwhile, the social sciences boomed, and new developments, such as environmental studies, became popular.

By the early 1970s, disillusion with the potential of liberal solutions to social and economic problems had set in, and the leaders saw the need for more radical social change, involving adoption of marxist, anarchist, and existentialist philosophies. Again, the social scientists were in the vanguard, but their growth potential was soon affected by the development of a recession in the economic fortunes of many western countries. To many observers, this recession was an inevitable consequence of the contradictions of capitalist economic growth and the unpreparedness of governments and business to tackle them; the exact timing was strongly affected by the Arab-Israeli Yom Kippur War of 1973 and the consequent use of oil as an economic weapon by the Arab countries. The price of this major industrial raw material increased rapidly and the economic problems of the dependent countries in Europe, and increasingly in North America too, were exacerbated. The expansion in universities was halted, in advance of any likely fall in demand for their services which might have come in the 1980s as a consequence of lower birth rates in the 1960s. Academia, after participating in and riding the boom of the previous decades, was thrown into depression.

Conclusions

The general theme of this book is that the recent history (indeed any history) of an academic discipline can only be understood in the context of its encompassing environment. This chapter has outlined the relevant environment for a study of Anglo-American human geography since 1945. This has involved the development of three arguments; one relating to how academic society works; another concerned with the organization of academic research; and the third with the economy and society which the academic disciplines serve. These three are, of course, complexly interrelated.

In the following chapters, the history of human geography since the Second World War will be outlined with an emphasis on what has been characterized here as 'extraordinary science', and less attention to the practice of 'normal science'. Those chapters are not organized to test any hypotheses which could be developed from the present set of arguments. The general tenor of the presentation is coloured by the contents of this chapter, however, and by the belief of some geographers that the models presented here, particularly that promoted by Kuhn, are relevant to an understanding of the changes that have occurred in the subject. The main purpose is to present a reasoned chronology, based on the published works of human geographers. The relationship between this chronology and models of academic disciplines only resurfaces as a major issue in the concluding chapter.

2

Foundations

Although this book is about human geography since 1945, the discussion of that period must be preceded by a brief outline of the nature of the discipline in the previous decades. Such a foundation is needed for a variety of reasons. The first is that although 1945 was something of a watershed year in many aspects of the social, economic and intellectual life of the countries being considered here, it did not mark a major divide in the views on geographical philosophy and methodology. Not surprisingly, the war years were not a major period of intra-discipline academic debate. Most academics spent much of the war either on active service or in associated intelligence activities (some of those involved in the latter retained their teaching commitments); the everyday activities of teaching, pure research, and administration were replaced by commitment to the war effort. It took a few years for academic life to return to something like normality, to assimilate the large numbers of new staff needed to replace the losses of the war years and to teach the backlog of students, and to react to the new social and economic environments.

A second reason for surveying the period preceding that being studied relates to the processes of change in academic work. New paradigms are created as reactions to those currently in favour, and not as inventions in an intellectual vacuum. Thus the post-Second World War changes were reactions to the philosophies and methodologies which had developed and been taught in the preceding decades; the nature of the reactions cannot be studied without some knowledge of what went before.

Finally, academic revolutions are not instantaneous affairs. A new paradigm usually takes years to mature, while experimentation with alternatives takes place, the programmatic statements are written, and the converts are won over by the prophets of the new approach. Meanwhile the current paradigm continues (or paradigms, if there are several with considerable support). Its adherents continue to work in their accepted ways, conducting research, publishing, and teaching generations of undergraduates according to the conventional wisdom. Even when a new paradigm has been crystallized, it may be that it must coexist for several years with its predecessor, as a competitor for the support of academics and students. This may be especially characteristic of the social sciences, in which interpretations of data are frequently more subjective than is the case with the physical sciences; it is quite feasible for two or more separate world views to find adherents at the same time, even in the same academic department.

Geography in the modern period

The hallmark of an academic discipline, according to one of geography's chroniclers (James, 1972), is that it has an educational organization which provides a specialist training in the subject. He dates the beginning of such an organization for geography at around 1874, when the first university geography departments were established in Germany: Britain and the United States followed a little later, with the main developments coming in the twentieth century. Before 1874, geography was a subject investigated either by amateurs or by scientists trained in other fields. With its own specialized institutional training, geography left its classical age and entered what James terms its modern period, which lasted for about eighty years, being superseded after 1945 by what James calls its contemporary period.

James's modern period is virtually co-extensive with the decades surveyed by Freeman (1961) in his *A Hundred Years of Geography* which, with James's book, is one of the few relatively recent attempts to provide a history of the discipline. Freeman identified six main trends in the geographical literature.

1 *The encyclopaedic trend*, associated with the collection of new information about the world, particularly areas little known to the residents of western Europe and North America. Although the great age of discovery was over, and by the late nineteenth century much of the world had been visited by Anglo-Saxon explorers, there were still vast tracts, notably in Africa, which if not *terra incognita* were extremely empty on contemporary maps. Indeed, at the beginning of geography's modern period much of the North American continent itself remained to be settled by permanent farms.

2 *The educational trend*, which, as James stresses, characterizes an academic discipline needing to propagate its knowledge, establish its relevance, and ensure its reproduction. Much work was undertaken to achieve a solid foundation of geographical work in schools, colleges and universities, involving both proselytizers and the architects of curricula.

3 *The colonial trend* reflects a major environmental preoccupation during the early decades of the modern period, especially in Britain whose empire was being consolidated and developed into a spatial division of labour based on the metropolitan hub and covering a considerable proportion of the earth's surface. Organization of the commercial world required a great deal of information about the various countries concerned, the provision of which became a major task of geographical research whilst its propagation was the keystone of geographical education.

4 *The generalizing trend* describes the use to which data collected in the encyclopaedic and colonial traditions were increasingly employed. Academic study involved more than the collection and collation of facts: these had to be interpreted, and the methods and aims of such interpretation defined the early paradigms of the discipline's development.

5 *The political trend* was reflected in the contemporary uses made of geographical expertise. Isaiah Bowman was a chief adviser to Woodrow Wilson at the conferences which re-drew the map of the world after the First World War, for example, and the work of geopoliticians such as Haushofer was influential on the *lebensraum* ideology of Nazi Germany.

6 *The specialization trend* was a reaction to the growth of knowledge and the inability of any one individual to master it all, even within the single discipline of geography.

Prior to the modern period, it was possible for scientists and other academics to be extremely catholic in their interests and expertise, but as the volume of research literature increased and the techniques of investigation demanded longer and more rigorous training so it became necessary for the individual to specialize, first as a geographer, and then within geography, focusing either on one substantive area or on a particular region of the earth's surface. Some of these trends represent philosophies, some methodologies, and some ideologies with regard to the purpose of academic geography. From them it is possible to identify three paradigms which characterized the modern period, both of human geography as a whole and of its component parts, such as urban geography (Herbert and Johnston, 1978). Discussion of these three occupies the remainder of this chapter.

Exploration

The first of the paradigms was carried over into the modern period from the classical, for exploration was the major activity recognized as geography through most of the nineteenth century. The collection and classification of information about 'unknown' parts of the earth (unknown, that is, to western Europeans and North Americans) was undertaken by geographers. Many of their expeditions were financed through the geographical societies which were founded during that century (Freeman, 1961); these, in turn, obtained money from commercial as well as philanthropical sources, for the information gathered was of great value to the mercantile world. As well as supporting and sponsoring exploratory expeditions, the geographical societies also undertook major educational roles. Their lecture meetings provided opportunities for the general public to see and hear of the new discoveries, and their officers worked hard to establish the teaching of geography in schools and universities: the Royal Geographical Society of London, for example, was involved in discussions which led to the establishment of geography teaching at England's two oldest universities, Oxford and Cambridge (Stoddart, 1975a).

The importance of exploration declined as geography matured in its new academic-discipline status during the early twentieth century, although in 1899 Halford Mackinder felt it necessary to establish his credentials as a geographer by becoming the first person to climb Mt Kenya. There was still much *terra incognita*, however, and the geographical societies maintained their interest in and sponsorship of expeditions. Indeed, the Royal Geographical Society still acts as a major sponsor of scientific expeditions large and small and its major publication—the *Geographical Journal*—reports the findings of many of them. The nature of the work undertaken is often very different from that of a century ago, reflecting developments in scientific technology and the available store of knowledge, but basic activities such as accurate map-making are still crucial to many successful expeditions. Many of the Society's meetings still present, in words and pictures, the results of expeditions and travels to all parts of the world, meetings which remain very popular with its large membership that is dominated by non-academics.

The American Geographical Society in New York is the American counterpart of the RGS. It has probably retreated a little further from the exploration role which it too focused on early in the modern period, and it now sponsors research in many other areas of geography, while still providing major support for investigations of relatively unstudied parts of the globe, such as the Arctic. The exploration tradition is maintained in the United States by the National Geographical Society and its popular journal, the *National Geographical Magazine*. Other societies have been established to take over some of the other professional roles: there are separate academic bodies in both the United States and Great Britain (the Association of American Geographers and the Institute of British Geographers) as well as those which concentrate on geographical education (the Geographical Association and the National Council for Geographical Education).

Although most of it was not strictly exploration, the work summarized by Freeman under the colonial trend can also be classified within this paradigm, since its aims were the collection, collation and dissemination of information. Much of the material put together was about commercial activities and infrastructure, as in volumes such as Chisholm's *Handbook of Commercial Geography* (first edition, 1899) and *Gazetteer of the World* (1895), which were aimed at the world of commerce, with companion volumes for schools. Their content comprised statistics and descriptions of production and trade, and a training in this type of geography was boring to many with its focus on the assimilation of large bodies of factual knowledge ('capes and bays' geography). But the existence of this geographical expertise was widely recognized, and was called on at the end of the modern period when geographers were made responsible for the preparation of intelligence reports about areas in which allied troops were likely to be engaged, work characterized by the set of British Admiralty Handbooks.

Environmental determinism and possibilism

These competing paradigms represent the first attempts at generalization by geographers of the modern period. Instead of merely presenting information in an organized manner, either topically or by area, geographers began to seek explanations for the patterns of human occupation of the earth's surface. The major initial source of their explanations was the physical environment, and a theoretical position was established around the belief that the nature of human activity was controlled by the parameters of the physical world within which it was set.

The origins of this environmental determinism lie in the work of Charles Darwin, whose seminal book *On the Origin of Species* (first published in 1859) influenced many scientists. His notions regarding evolution were taken up by the American geographer William Morris Davis in his famous cycle-of-erosion model of landform development. Ideas of natural selection and adaptation formed the basis of statements regarding environmental determinism, including one by Davis (1906) whose programmatic paper identified the core of geography as the relationship between the physical environment, the control, and human behaviour, the response (Stoddart, 1966).

Chief among the early nineteenth-century environmental determinists was the German geographer Ratzel, whose American disciple Ellen Churchill Semple opened her book *Influences of Geographic Environment* (1911) with the statement that 'Man is the product of the earth's surface'. In some hands, the environmental influences adduced were gross, and with hindsight it is hard to believe that they could have been written and taken seriously; a brief survey by Tatham (1953), for example, illustrates the extent to which authors were prepared to credit all aspects of human behaviour with an environmental cause.

Reaction to the extreme generalizations of the environmental determinists led to the development of a counter-thesis, that of possibilism, in which man was presented as the active rather than the passive agent. Led by French geographers, followers of the historian Lucien Febvre, possibilists presented a model of man perceiving the range of alternative uses to which he could put an environment and selecting that which best fitted his cultural dispositions. Taken to extremes, this paradigm could be as ludicrous as that which it opposed, but in general the possibilists recognized the limits to action which environments set, and avoided the great generalizations which characterized their antagonists.

Debate over environmental determinism and possibilism continued into the 1960s (Lewthwaite, 1966) and was actively pursued in the United Kingdom in the first decade after the Second World War. The determinism cause was continued in the period between the world wars by writers such as Ellsworth Huntington, who advanced theories relating the course of civilization to climate and climatic change. Perhaps the most doughty of all advocates was the Australian Griffith Taylor, whose determinist views so angered politicians interested in the settlement of outback Australia that he was virtually hounded out of his homeland. Taylor's reaction to Tatham's (1953, p. 150) statement that when industrialists decide where to locate a new factory 'Geographical controls are rarely mentioned' was, 'Surely this definitely illustrates the stupidity of the owner!' His case was that the possibilists had developed their arguments in temperate environments such as that of northwestern Europe, which do indeed offer several viable alternative forms of human occupance. But such environments are rare: in most of the world—as in Australia—the environment is much more extreme and its control over man that much greater accordingly He coined the phrase 'stop-and-go' determinism to describe his views. In the short term, man might attempt whatever he wished with regard to his environment, but in the long term, nature's plan would ensure that the environment won the battle and forced a compromise out of its human occupants.

Many debates begin by being based on two opposing, extreme positions, and end as a compromise accepted by all the most fervent devotees of either polar position. Thus the lengthy discussions among geographers about whether man is a free agent in his use of the earth, or whether there is a 'nature's plan', slowly dissolved as the antagonists realized the existence of merits in each case. But while environmental determinism was a view strongly held and widely preached by geographers seeking a niche within the halls of science, the relative standing of the discipline declined somewhat in the eyes of the academic community at large, who rejected the paradigm. As a consequence, geography's next paradigm, which had strong roots in environmental determinism, was very much an introspective and conservative one.

The region and regionalism

This, the third of the paradigms, dominated British and American geography for much of the first half of the present century. Like environmental determinism, it too was an attempt at generalization, but at generalization without structured explanation, and thus of a very different type from the increasingly discredited law-making attempts of the previous paradigm. Much of the early development took place in Britain, and involved work at two scales (Freeman, 1961, p. 84). At the large scale were efforts, such as Herbertson's (1905), to divide the earth into major natural regions, usually on the basis of climatic parameters and thus having some links with the earlier determinism. At the smaller scale, the aim was to identify individual areas with particular characters:

> The fundamental idea was that the small area would legitimately be expected to show some distinct individuality, if not necessarily entire homogeneity, through a study of *all* its geographical features—structure, climate, soils, vegetation, agriculture, mineral and industrial resources, communications, settlement and distribution of population. All these, it has often been said, are united in the visible landscape, linked into one whole and dependent one on another. And more, every area, save those few never occupied by man, has been influenced, developed and altered by human activity, and therefore the landscape is an end-product, moulded to its present aspect by successive generations of people. The practice has therefore been to take an evolutionary view and . . . to attempt to reconstruct the landscape as it was a hundred, or a thousand years ago (Freeman, 1961, p. 85).

Hartshorne and the American view

The ideas and methods of regional geography were taken up a little later in the United States. In the late 1930s, however, two non-geographers published a major survey of American regionalism (Odum and Moore, 1938) and in 1939 the Association of American Geographers published a monograph—Richard Hartshorne's *The Nature of Geography: A Critical Survey of Current Thought in the Light of the Past*—which rapidly established itself as the definitive statement of the paradigm. As Hartshorne (1948) later made clear, there was much debate among American geographers during the 1930s (most of it apparently unpublished) about the nature of their discipline. Hartshorne was concerned about both tone and content of the debate, and in 1938 submitted a paper to the *Annals*, as a contribution to the philosophical discussions. He then proceeded to Europe for fieldwork on boundary problems, as part of his ongoing research into political geography. This work was frustrated by the political situation, and so he spent his time reading European, mainly German, work on the nature of geography. He used this to extend his 1938 paper, adding the sub-title; the result was a 'paper' of 491 pages (some 230,000 words) which became the major philosophical and methodological contribution to the literature of geography in English then available.

A synopsis of Hartshorne's scholarship, and his interpretations of the scholarship of others, notably Hettner, is not possible in a few paragraphs, and only the main conclusions can be stressed here. Hartshorne argued forcefully that the focus of geography is areal differentiation, the mosaic of separate landscape on the earth's surface. Thus the discipline is:

> a science that interprets the realities of areal differentiation of the world as they are found, not only in terms of the differences in certain things from place to place, but also in terms of the total combination of phenomena in each place, different from those at every other place (p. 462)

so that

geography is concerned to provide accurate, orderly and rational description and interpretation of the variable character of the earth surface (p. 21)

and it

seeks to acquire a complete knowledge of the areal differentiation of the world, and therefore discriminates among the phenomena that vary in different parts of the world only in terms of their geographic significance—i.e. their relation to the total differentiation of areas. Phenomena significant to areal differentiation have areal expression—not necessarily in terms of physical extent over the ground, but as a characteristic of an area of more or less definite extent (p. 463).

According to this view, the principal purpose of geographical scholarship is synthesis, an integration of relevant characteristics to provide a total description of a place—a region—which is identifiable by its peculiar combination of those characteristics. There is then, according to Hartshorne, a close analogy between geography and history; the latter provides a synthesis for 'temporal sections of reality' whereas the former performs a similar task for 'spatial sections of the earth's surface' (p. 460).

Hartshorne also indicated the methodology to be used for this integrating science aimed at orderly description of the earth's surface. To him, 'the ultimate purpose of geography, the study of areal differentiation of the world, is most clearly expressed in regional geography' and so accepted procedures were necessary for regional identification. Regions are characterized by their homogeneity on prescribed characteristics, selected for their salience in highlighting areal differences. Two types of region were identified; the *formal* region (or uniform region) in which the whole of the area is homogeneous with regard to the phenomenon or phenomena under review, and the nodal or *functional* region in which the unity is imparted by organization around a common node, which may be the core area of a state or a town at the centre of a trade area. Identification of such regions

depends first and fundamentally on the comparison of maps depicting the areal expression of individual phenomena, or of interrelated phenomena . . . geography is represented in the world of knowledge primarily by its technique of map use (pp. 462–4).

Hartshorne placed his emphasis on map *use*. Although it is valuable for geographers to know something about the preparation and construction of maps, the sciences of surveying and map projections are of only secondary interest to them; the prime task of the geographer is in the interpretation of maps, and increasingly, from about 1940 on, of various forms of aerial photograph. Much of this information to be interpreted may have been placed on the maps by geographers during their fieldwork, and the role and nature of fieldwork were of considerable interest to American geographers during the period when Hartshorne was developing his ideas.

Preparation of the materials for a regional synthesis required collection both from other sciences specializing in certain phenomena (though usually not their areal patterning) and from the topical systematic specialisms which complemented, but which were eventually subsidiary to, regional geography. Physical, economic, historical and political were the main systematic subdivisions recognized within geography at the time Hartshorne wrote, although a later survey, set firmly within the regional paradigm, identified many other 'adjectival geographies', including population, settlement, urban, resources, marketing, recreation, agricultural, mineral production, manufacturing, transportation, climatology, geomorphology, soils, plant, animal, medical, and military (James and Jones, 1954). A number of these were of only minor importance, however, so that despite the apparent diversity of interests among geographers of the time, the 'classic' regional study usually followed a sequence comprising physical features, climate, vegetation, agriculture, industries, population and the like (Freeman, 1961, p. 142) and summarized by a synthesis of the individual maps to produce a set of formal regions.

To most geographers of the period spanning World War II, and notably those who contributed to the survey edited by James and Jones (1954), regional geography was at the forefront of their discipline's scholarship and systematic studies were the providers of information for that enterprise: thus to James, 'Regional geography in the traditional sense seeks to bring together in an areal setting various matters which are treated separately in topical geography' (1954, p. 9). Urban geographers studied towns because they 'constitute distinctive areas' (Mayer, 1954, p. 143), in line with the regional concept; political geographers studied the functions and structures of an area 'as a region homogeneous in political organization, heterogeneous in other respects' (Hartshorne, 1954, p. 174); and in defining the 'new' field of social geography, Watson (1953, p. 482) saw it 'as the identification of different regions of the earth's surface according to associations of social phenomena related to the total environment'. Each of these topical specialisms, therefore, sought to produce its own regionalization (notable in this was the work of agricultural geographers, especially O. E. Baker, in a series of papers published in *Economic Geography* during the 1920s and 1930s outlining the agricultural regions of various parts of the world). It had its links with the relevant systematic sciences—social geography with sociology, for example—and the key differentiating factor between the two was the geographer's focus on the region, the single-attribute region of his specialism and the multi-attribute region in the synthesis of his work with that of others to produce regional geographies.

Given this focus on the region, it is not surprising that the literature of the paradigm contained many contributions discussing the nature and delimitation of such homogeneous areas, for virtually every region was in effect a generalization, complete homogeneity being very rare over more than a small area. As already indicated, British geographers were active early in the definition of large-scale regions, usually based on climatic parameters. Much effort was made to develop methods to define multi-attribute regions; in agricultural geography, for example, it culminated in the statistical method developed by Weaver (1954). But at the small scale it was widely accepted that regional delimitation should be based on personal interpretation of landscape assemblages. For this, the model was the work by the French geographer Paul Vidal de la Blache and his followers on the *pays* of their homeland, small regional units with distinct physical characteristics, notably in soils and drainage, and associated agricultural specialisms (Buttimer, 1971, 1978a).

One systematic specialism which stood slightly apart from the others was historical geography, the study of which was based on the argument that investigations of genesis were needed in order to comprehend the regional patterns of the present. Two approaches to historical geography can be recognized in its literature from the 1920s on. The first, often thought of as the British approach and closely associated with the work of H. C. Darby, involved the detailed study of past geographies (Perry, 1969). This was done in a series of cross-sections, whose locations in time were almost always determined by the available source material, such as the Domesday Book of *c.* 1086 which was analysed in great depth by Darby and his associates (culminating in Darby, 1977). These cross-sectional analyses, complete with their regionalizations in many cases, were linked together by a narrative outlining the changes between the periods studied: most emphasis was placed on the cross-sections, however, for which data allowed analysis rather than interpretation (see Darby, 1973).

The second approach was largely American in its provenance, and centred upon the works of C. O. Sauer and his associates. The focus was the study of ongoing processes leading to landscape change up to, and including, the present and beginning at the prehuman stage of occupance (Mikesell, 1969): most of the work was conducted either outside the United States itself (particularly in Latin America) or in the less industrialized parts of that country. Sauer's (1925) first methodological statement constrained geographical endeavour closely to the generic study of landscapes, with emphasis on their cultural features (although in fact work was also done on the borderlands between geography and botany); there was no glorification of the region, however. In his later sermons—as he called his methodological and philosophical statements—Sauer (1941, 1956) encouraged research over a much wider field, but emphasized the study of cultural landscapes and the links which he had forged with anthropology, to produce a creative art-form whose hallmark was that it was not prescribed by pattern or method: the human geographer is obliged 'to make cultural processes the base of his thinking and observation' (Sauer, 1941, p. 24). The work, as undertaken by Sauer and his students, involved neither detailed reconstruction of past geographies nor close consideration of regional boundaries: instead it led to a catholic historical geography whose rationale (Clarke, 1954, p. 95) was that:

> through its study we may be able to find more complete and better answers to the problems of interpretation of the world both as it is now and as it has been at different times in the past.

Not all American historical geographers followed this lead—Brown (1943), for example, worked on detailed reconstructions of past periods—but the 'Berkeley School' which Sauer founded and led for almost five decades had many followers and a point of view within the prevailing academic orientation of geography which almost made it a separate paradigm within a paradigm, focused on a single iconoclast.

The British view
British geographers seem to have been less concerned with philosophical and methodological debate than their American counterparts during the 1920s, 1930s, and 1940s (though see the exchange in the *Scottish Geographical Magazine* during the late 1930s, initiated by Crowe, 1938). They seem to have been more pragmatic in their work, less prone to contemplate the nature of their subject and more prepared, perhaps, to adopt the well used adage that 'Geography is what geographers do'. But they too accepted that the *raison d'être* of geography was synthesis, the integration of the findings of various systematic studies with a strong emphasis on genesis, as in the studies of geomorphology and historical geography (Darby, 1953). According to Wooldridge and East (1958):

> geography . . . fuses the results, if not the methods, of a host of other subjects . . . [it] is not a science but merely an aggregate of sciences (p. 14)
> its *raison d'être* and intellectual attraction arise in large part from the shortcomings of the uncoordinated intellectual world bequeathed us by the specialists (pp. 25–6)
> in its simplest essence the geographical problem is how and why does one part of the earth's surface differ from another (p. 28)

all of which statements indicate a strong trans-Atlantic common body of opinion, although, despite a statement that 'The purpose of regional geography is simply the better understanding of a complex whole by the study of its constituent parts' (p. 159), the British writers did not elevate the regional doctrine as much as did their American counterparts. Nevertheless, Wooldridge (1956, p. 53) wrote in 1951 that the aim of regional geography . . . is to gather up the disparate strands of the systematic studies, the geographical aspects of other disciplines, into a coherent and focused unity, to see nature and nurture, physique and personality as closely related and interdependent elements in specific regions

and that in any department of geography each member of the staff should commit himself to the study of a major region (p. 64).

One major difference between British and American geography by the 1950s was in their attitudes to physical geography, the study of the land surface, the atmosphere and the oceans, and their faunal and floral inhabitants. Both countries had strong traditions of work on these topics, and many geographers had academic roots in the associated field of geology. But in North America (the United States much more than Canada) this tradition had slowly dissolved and interest in the physical environment, and particularly its understanding as against its description, waned. This may have been a consequence of the excesses of environmental determinism, and a subsequent desire to remove all traces of that connection and to see man as the formative agent of landscape patterns and change: associated with this was probably the attempt to redefine geography in the 1920s as the study of human ecology, in which man is seen as reacting and adjusting to environments while at the same time attempting to adjust the environment to his own needs (Barrows, 1923). Thus with regard to geomorphology—the science of landform genesis—Peltier (1954, p. 375) wrote:

the geographer needs precise, factual information about particular places. What landforms actually exist in a given area? How do they differ? Where are they? What are their distribution patterns? The geomorphologist may concern himself with questions of structure, process, and stage, but the geographer wants specific answers to the questions: what? where? and how much?

What geographers were interested in, according to this view, was the geography of landforms: geomorphology, the genetic study of landforms, was a part of geology and, unlike historical geography, was deemed irrelevant to the geographical enterprise. Similar reactions saw the wholesale removal of climatology and biogeography from American geographical curricula, and their replacement by introductory courses in physical geography which described landforms, climates, and plant assemblages—usually in a regional context—but paid little or no attention to their origins.

This American trend was not repeated in Britain where, according to Wooldridge and East (1958, p. 47):

To treat geography too literally as an affair of the 'quasistatic present' is to make both it and its students seem foolish and superficial. It is true that our primary aim is to describe the present landscape; but it is also to interpret it. . . . Our study has therefore always to be evolutionary. . . . It is unscholarly to take either landforms or human societies as 'given' and static facts, though we must not let temporal sequences obscure spatial patterns.

Thus, certainly in the 1950s, geography students at British universities rarely specialized, except perhaps in the final year of their course, in either physical or human geography. Both were considered essential parts of a geographical education, as contributions to the genetic study of regional landscapes which was the integrating focus of geographical scholarship. As researchers, most British geographers specialized in either physical or human geography (though rarely exclusively so), but almost all had a regional specialism as well, in which they integrated studies from 'both sides' of their subject, as widely illustrated in the regional textbooks of the period.

Conclusions

This chapter has presented an extremely brief outline of geography during its 'modern period', since the focus of the book is on the ensuing 'contemporary period'. Three paradigms have been identified, although deeper analysis may well indicate more coexisting during any one time (see Taylor, 1937). All three lasted into the contemporary period, although one, the regional paradigm, was dominant in the years before and just after the Second World War. Its main focus was on areal differentiation, on the varying character of the earth's surface (basically, the inhabited parts of the earth's surface), and its picture of that variation was built up out of parallel topical studies of different aspects of the physical and human patterns observed. By the 1950s, initially in America and then in Britain too, disillusionment with the philosophy of the regional paradigm was growing. Slowly the topical specialisms came to greater dominance and the regional synthesis was ignored: eventually, as outlined in the next chapter, a full revolution was launched with the presentation and ultimate wide acceptance of a new paradigm.

3

Growth of systematic studies and the adoption of 'scientific method'

Dating the origin of a change in the orientation of a discipline, or even a part of it, is difficult. Several pieces which contain the kernel of the new ideas can usually be found in its literature, but often these are derivative of the earlier teachings of others, whose views are never published. Further, it is possible for a change to emanate contemporaneously from several separate though usually not entirely independent nodes, as several iconoclasts introduce stimuli to change. An attempt to locate the first stirrings against the regional paradigm in human geography would be a futile exercise, therefore, and is not attempted here. Instead, the present chapter isolates what appear to have been the most important and influential statements published by geographers, and traces their impact on the geographical community of scholars.

As pointed out in Chapter 1, a revolution involves both dissatisfaction with the existing paradigm(s) and the preparation of an acceptable alternative. The existence of the former was spelled out by Freeman (1961), who noted that 'disapointment with the work of regional geographers has led many to wonder if the regional approach can ever be academically satisfying and to turn to specialization or some systematic branch of the subject' (p. 141). He suggested three reasons for such disappointment. The first was that so much regional classification was naive, particularly at the large scale, where generalizations, such as Herbertson's world climatic regions, were found on detailed investigation to contain too many discrepancies. The second, and perhaps most important to many people, was the 'weary succession' of physical and human activity 'facts' which characterized so much regional writing (though not all, as exemplified by James's, 1942, *Latin America*): 'The trouble has perhaps been that many regional geographers have tried to include too much' (p. 143). Thirdly, he claimed that the model of regional writing, derived from work on the French *pays*, suggested that the whole of the earth's surface could be divided into such clear regions, each with its own character: that this proved to be not so was reflected by many pedestrian studies of areas lacking such personality.

Whereas Freeman focused on the failings of regional geography as practised, a case made in the United States during the late 1940s and early 1950s was that the insistence of the primacy of regional geography was undermining the associated systematic studies. This was put force-fully by Ackerman (1945) in a paper reporting on his experience of working in the wartime intelligence services. He identified two major failings of professional geographers involved in such activities: their inability to handle foreign languages, and the weakness of their topical specialisms. Regarding the latter, he criticized much of the geographical work of the preceding quarter of a century as having been conducted by scholars who were 'more or less amateurs in the subject on which they published' (p. 124), so that when they were called upon to provide intelligence material for wartime interpretation what the geographers produced was extremely thin in its content. Regional geographers could provide no more than a superficial analysis, and the division of labour within the discipline whereby people specialized on different areas of the earth was both inefficient and ineffective.

Ackerman suggested that rectification of this major deficiency in geographical work required much more research and training in the systematic specialisms: this would not be contrary to the philosophy of the subject which gave primacy to the regional synthesis, he claimed, since more detailed systematic studies would lead to greater depth in regional interpretations. There is little evidence that his paper had an immediate impact, however, and the publications of American geog-raphers over the next few years, including the abstracts of the papers presented at the annual conferences of the Association of American Geographers, indicated no major shift in the orientation of academic work with the return to post-war 'normality': one of the few exceptions to this is the abstract of a paper presented by Garrison at Cleveland in 1953, which was clearly based on a different methodology to that widely used (see below, p. 54). The systematic fields had undoubtedly been gaining in importance prior to Ackerman's statement, and continued to do so, as indicated by the extent of their treatment in the review volume edited by James and Jones (1954). But it was not until the mid-1950s that this volumetric change in the substance of geographical research was matched by any widespread changes in its methodology and philosophy.

Schaefer's paper and the response

As it was in the United States that Hartshorne published his major statement of the regional paradigm, and as it was there, rather than in Britain, that philosophy and methodology were apparently debated most earnestly, it is perhaps not surprising that the revolution against the regional paradigm originated on that side of the Atlantic. One of the first shots in that revolution was a paper by Schaefer (1953)—which was published posthumously—that is often referred to by those who seek the origins of the 'quantitative and theoretical revolutions'. Schaefer was originally an economist: he joined the group of geographers teaching in the economics department at the University of Iowa after his escape from Nazi Germany.

Schaefer pointed out that his paper was the first to challenge Hartshorne's presentation and interpretation of the works of Hettner and others, and it was published fourteen years after Hartshorne's monograph. His intent was to criticize the 'exceptionalist' claims made for regional geography, and to present the case for geography adopting the philosophy and methods of the logical-positivist school of science. His first task, then, was to outline the nature of a science and to define the peculiar characteristics of geography as a social science. He argued that to claim that geography was the integrating science which put together the findings of the individual systematic sciences was arrogant, and that in any case its products were 'somewhat lacking in . . . startlingly new and deeper insights' (p. 227). A science is characterized by its explanations, and explanations require laws:

> To explain the phenomena one has described means always to recognize them as instances of laws (p. 227).

In geography, according to Schaefer, the major regularities which are described refer to spatial patterns:

> Hence geography has to be conceived as the science concerned with the formulation of the laws governing the spatial distribution of certain features on the surface of the earth (p. 227)

and it is these spatial arrangements of phenomena, and not the phenomena themselves, about which geographers should be seeking to make law-like statements. Geographical procedures would then not differ from those employed in the other sciences, both natural and social: observation would lead to a hypothesis—about the interrelationship between two spatial patterns, for example—and this would be tested against large numbers of cases, to provide the material for a law if it were thereby verified.

The argument against this definition of geography as the science of spatial arrangements—the implicit argument in Hartshorne's work— Schaefer termed exceptionalist. It claims that geography does not share the methodology of other sciences because of the peculiar nature of its subject matter—the study of unique places, or regions. Using analogies from physics and economics, Schaefer argued that geography is not peculiar in its focus on unique phenomena; all sciences deal with unique events which can only be accounted for by an integration of laws from various systematic sciences, but this does not prevent—although it undoubtedly makes more difficult—the development of those laws.

> It is, therefore, absurd to maintain that the geographers are distinguished among the scientists through the integration of heterogeneous phenomena

which they achieve. There is nothing extraordinary about geography in that respect (p. 231).

In the second part of his paper, Schaefer traced the exceptionalist view in geography back to an analogy drawn by Kant (1724–1804) between geography and history, an analogy repeated by Hettner and by Hartshorne (see above, p. 35). He quoted (p. 233) from Kant's *Physische Geographie* (Vol. I, p. 8) that 'Geography and history together fill up the entire area of our perception: geography that of space and history that of time.' But when Kant was working, Schaefer claims, history and geography were cosmologies, not sciences, and a cosmology is 'not rational science but at best thoughtful contemplation of the universe' (p. 332). Hettner, however, followed Kant's views and developed geography as a cosmology, arguing that both history and geography deal with the unique, and thus do not apply the methods of science. Schaefer argued that this is a false position, for in explaining what happened at a certain time period historians must integrate the laws of the social sciences. Time periods, like places, are undoubtedly unique assemblages of phenomena, but this does not preclude the use of laws in unravelling and explaining them. History and geography can both be sciences for

> What scientists do is . . . *They apply to each concrete situation jointly all the laws that involve the variables they have reason to believe are relevant* (p. 239).

Hartshorne argued that law-seeking is not a part of geography. According to Schaefer, however, Hartshorne disregarded one aspect of Hettner's writing which was nomothetic in its orientation, and in doing this he to some extent misled American geographers.

The final part of Schaefer's paper reviews some of the problems of applying his nomothetic (law-producing) philosophy to geography as a spatial, social science. He recognizes, for example, the problems of experimentation and of quantification, and suggests a methodology based on cartographic correlations. A major point concerns the difference between laws produced in geography and those from other, 'maturer' social sciences. The former are morphological; the latter are process: in order fully to comprehend the assemblages of the phenomena described in geographers' morphological laws, therefore, it is necessary to derive process laws from other social sciences, a procedure which requires team work (the last point was made also by Ackerman). Geography, then, according to Schaefer, is the source of the laws on location, which may be used to differentiate the regions of the earth's surface.

Hartshorne's response
Schaefer's paper did not produce much direct reaction, in print, despite later claims that it was a major stimulus to work in the genre which he proposed (Bunge, 1962). It did, however, draw considerable response

from Hartshorne, in the form first of a letter to the editor of the *Annals* (Hartshorne, 1954b) and later three substantive pieces (Hartshorne, 1955, 1958, 1959): the last of the latter was another major book which, although probably not as influential as the 1939 volume, showed the continued importance of Hartshorne to American geographers as an interpreter of their subject's methodology and philosophy.

The purpose of Hartshorne's (1955) paper (which subsumed the earlier letter) was to indicate the many flaws which he identified in Schaefer's scholarship (see also Gregory, 1978a, p. 31). He begins with a further discussion of the *mores* of methodological debate (Hartshorne, 1948): most of the paper was organized to illustrate that Schaefer was limited in his references, drew unsupportable conclusions, and misrepresented the views of others, so that 'In every paragraph, in nearly every sentence of this third section, there is serious falsification, either by commission or by omission, of the views of the writer discussed' (p. 236). (It should be noted that this statement refers to the third part of Schaefer's paper, which focused on the work of Hettner.) In more general terms Hartshorne claims that Schaefer's paper 'ignores the normal standards of critical scholarship and in effect offers nothing more than personal opinion, thinly disguised as literary and historical analysis' (p. 224). Since Hartshorne himself (1959, p. 8) is a strong believer that 'geography is what geographers have made it', to him all methodological and philosophical statements should be based on a close and careful analysis of the published works of others.

Although most of this paper is concerned to examine the nature of Schaefer's 'evidence', in the final section Hartshorne turns to an examination of the anti-exceptionalist argument. He points out (p. 237) that in coming to the conclusion that geography should take process laws from the systematic sciences and use then to produce morphological laws, Schaefer came very close in effect to preaching the sort of exceptionalist claim that he sought to destroy. It could be argued, therefore, that Schaefer's critique 'is a total fraud' (p. 237). Schaefer's position is summed up as 'geography must be a science, science is the search for laws, and all phenomena of nature and human life are subject to such laws and completely determinable by them' (p. 242). Such scientific determinism is opposed to the summary of what geographers do as set out in *The Nature of Geography*, which has in any case been treated in a most cavalier way by Schaefer.

In his second paper, Hartshorne (1958) addressed Schaefer's claim that Kant was the source of the exceptionalist view. Literary analysis suggests that both Humbolt and Hettner reached the same view independently, being unaware of Kant's views when they were writing. May (1970, p. 9) suggests that both Hartshorne and Schaefer could have misunderstood Kant's conception of a science, however, and of the role of geography as a science, although he confirms Hartshorne's dismissal of Schaefer's interpretation of the source of Kant's ideas (see the later exchange between Hartshorne, 1972 and May, 1972).

The third and most substantial piece in Hartshorne's rebuttal of Schaefer's argument was a monograph (Hartshorne, 1959) entitled *Perspective on the Nature of Geography*, the production of which was stimulated by Schaefer—and by requests from colleagues that he respond in detail to Schaefer's argument—but which was also used as a vehicle for a discussion of a wide range of other issues raised during the two decades since the publication of his original statement (Hartshorne, 1939). He organized the discussion in a framework of ten separate questions/topics: the aim was to provide a methodology by which geography could meet its need for 'new conceptual approaches and more effective ways of measuring the interrelationships of phenomena' (p. 9), which could only develop out of an understanding and acceptance of the subject's 'essential character'.

The first set of questions was concerned with the meaning of areal differentiation, with the definition of the earth's surface, with a discussion of the peculiar geographical interest in the integration of phenomena in 'the total reality [that] is there for study, and geography is the name of the section of empirical knowledge which has always been called upon to study that reality' (p. 33), and with the determination of what is significant for geographical study; it led Hartshorne to the definition that 'geography is that discipline that seeks *to describe and interpret the variable character from place to place of the earth as the world of man*' (p. 47). He considered that human and natural factors do not have to be identified separately—any prior insistence on this was a function of the arguments of environmental determinists—and that a division into human and physical geography is unfortunate, because it limits the range of possible integrations in the study of reality.

Turning to the study of temporal processes, Hartshorne argued that geographers need only study proximate genesis, since it is classification by form of appearance rather than by provenance which is important for the geographical investigation of areal differentiation: since most landforms are stable, or virtually so, from the point of view of man, for example, the study of their change is irrelevant to the aims of geography. According to this argument (see also p. 39), geomorphology, insofar as it is the study of landform genesis, is not part of geography; the study of landforms is. With regard to cultural features in the landscape, Hartshorne made an important distinction between expository description and explanatory description:

> geography is primarily concerned to describe . . . the variable character of areas as formed by existing features in interrelationships . . . explanatory description of features in the past must be kept subordinate to the primary purpose (p. 99).

Thus historical geography should be the expository description of the historical present 'but the purpose of such dips into the past is not to trace developments or seek origins but to facilitate comprehension of the present' (p. 106); studies of causal development and genesis are the prerogative of the systematic sciences.

In attempting an answer to the question 'Is geography divided between systematic and regional geography?' Hartshorne developed a position different from that which he took in *The Nature of Geography*. Thus in 1959 he accepted that studies of interrelationships could be arranged along a continuum, 'from those which analyse the most elementary complexes in areal variation over the world to those which analyse the most complex integrations in areal variation within small areas' (p. 121). The former are called topical studies and the latter regional studies, but whereas

> every truly geographical study involves the use of both the topical and the regional approach (p. 122)

there is no argument that one is superior over the other, as being that to which all geographers should aspire. In this presentation, therefore, Hartshorne somewhat downgraded the regional synthesis from its earlier centrality, in his view, in the geographical enterprise.

With regard to the important question raised by Schaefer's paper— 'Does geography seek to formulate scientific laws or to describe individual cases?'—Hartshorne argued for the latter, largely by pointing out the difficulties of establishing such laws through geographical investigations. Scientific laws must be based on large numbers of cases, but geographers study complex integrations in unique places; scientific laws can best be established in laboratory experiments which allow only a few independent variables to vary, but such work is impossible in geography; interpretation requires skills in the systematic sciences which are beyond the capability of geographers; scientific laws suggest some kind of determinism, but this is inappropriate to the human motivations which are in part the causes of landscape variations: for these reasons, the search for laws is irrelevant to geography. But laws are not the only means to the scientific end of comprehending reality in any case: instead

> Geography seeks (1) on the basis of empirical observation as independent as possible of the person of the observer, to describe phenomena with the maximum degree of accuracy and certainty; (2) on this basis, to classify the phenomena, as far as reality permits, in terms of generic concepts or universals; (3) through rational consideration of the facts thus secured and by logical processes of analysis and synthesis, including the construction and use wherever possible of general principles or laws of generic relationships, to attain the maximum comprehension of the scientific interrelationships of phenomena; and (4) to arrange these findings in orderly systems so that what is known leads directly to the margin of the unknown (pp. 169–70),

which, he says, is a perfectly respectable scientific goal.

Finally, in discussing geography's position within the classification of sciences, Hartshorne returned to the Hettnerian analogy of geography as a chorological science with history as a chronological science. This is valid, he argues, because it describes the way in which geographers have worked, on both topical and regional subjects, with reference to interrelationships and integrations within areas.

Reconciliations?

The last statement derived from Hartshorne in the previous paragraph indicates the major basis of the methodological and philosophical difference between himself and Schaefer. Hartshorne's was a positive view of geography: geography is what geographers have made it. Schaefer's view, on the other hand, was a normative one, of what geography should be, irrespective of what it had been. Over the next decade after Hartshorne published his *Perspective* it was Schaefer's view which very largely prevailed, on both sides of the Atlantic, although the extent of Schaefer's personal influence based on his 1953 paper was probably very slight and the real iconoclasts of the 'revolution' were those discussed in the next section. (Indeed in Britain, although Hartshorne's two books were clearly widely read and referenced, Schaefer's paper was not. It receives no mention in Freeman's (1961) book, none in Chorley and Haggett's (1965b) trail-blazing *Frontiers in Geographical Teaching*, and only one in their major (Chorley and Haggett, 1967) *Models in Geography*—in the chapter by Stoddart.) Thus it is not surprising that relatively little attention has been paid elsewhere in the geographical literature to the Schafer/Hartshorne debate (Gregory, 1978a, p. 32).

An attempt at reconciling the views of the two protagonists, and to suggest that they were not so antagonistic in their views as they themselves suggested, has been provided by Guelke (1977a, 1978; see also Gregory, 1978a, p. 31). He shows that in general terms Hartshorne was very much a supporter of the scientific method as defined by the logical positivists (see below, p. 62), but that he created his own problems regarding the application of this method in geography because of his view on uniqueness. Schaefer, on the other hand, accepted the full logical-positivist position, and showed that uniqueness was a general problem of science, and not a peculiar characteristic of geography, thus

> In extending the idea of uniqueness to everything, Schaefer effectively removed a major logical objection to the possibility of a law-seeking geography and demonstrated that Hartshorne's view of uniqueness as a special problem was untenable for anyone who accepted the scientific model of explanation (Guelke, 1977a, p. 380),

and Hartshorne's distinction between idiographic and nomothetic approaches was misleading. Both Hartshorne and Schaefer ignored the possibility of geographers being major 'law-consumers', however; to Hartshorne, the alternatives were either law-making or the description of unique places, whilst to Schaefer geographers had to develop morphological laws, and ignore the interest in process laws which characterizes the systematic sciences.

According to Guelke (1977a, p. 384), when Schaefer insisted on the need for geographers to develop laws 'he created a major crisis within the discipline'. Whether Schaefer himself was responsible for the crisis is doubtful, as the next section suggests. There is no doubt, however, that within about a decade of Schaefer's paper being published, most human geographers, and certainly those of the youngest generation currently within the profession, had adopted at least part of his manifesto, with the growing concern for quantification and law-making. They were presented with a choice between such activity and the sort of contemplation of the unique advocated by Hartshorne. As Guelke (1977a, p. 385) points out, 'Not surprisingly, most geographers opted for geography as a law-seeking science' because (Guelke, 1978, p. 45) by then:

> Universities were expected to produce problem-solvers or social-technologists to run increasingly complex economies, and geographers were not slow in adopting new positions appropriate to the new conditions. Statistics and models were ideal tools for monitoring and planning in complex industrial societies. The work of the new geographers, however, often lacked a truly intellectual dimension. Many geographers were asking: 'Are our methods rigorous?', 'What are the planning implications of this model?', and not 'How much insight does this study give us?', 'Is my understanding of this phenomenon enhanced?', 'Does this study contribute to geography?'. The last-mentioned question was considered of little consequence. Yet it should have been asked, because one of the weaknesses of the new geography was a lack of coherence.

Before analysing this last statement further, however, it is necessary to look in detail at what was done in the 'new geography' of the 1950s and 1960s.

Developments in systematic geography in the United States

Whether because of, or independently of, the statements of Ackerman (1945), Schaefer (1953), and Ullman (1953), it is clear that during the 1950s systematic studies became much more important in the research and teaching of American geographers. This did not mean a departure from Hartshorne's views, since by 1959 he no longer gave primacy to regional studies, but the trend towards the scientific method proposed by Schaefer did mark a break with the Hartshornian tradition.

The growing popularity of topical specialisms is shown by the review chapters in the collection edited by James and Jones (1954) and by the journal literature of the 1950s. Very few of the investigations reported were aimed at the generation of laws in any sense, however: indeed some could almost be categorized under the exploration paradigm, in that their major purpose seemed to be the provision of new factual material about, for example, the changing distribution of Soviet pulp and paper industries (Rodgers, 1955). Others can more appropriately be classified as regional studies, with foci on single phenomena. There was very little attempt at generalization.

Fundamental to the progress of science in the logical-positivist mould espoused by Schaefer is the development of theory. Several of the reviews in the James and Jones (1954a) volume refer to what is in one place termed location theory (Harris, 1954, p. 299), but very few examples are cited of empirical investigations related to that body of theory. The chapter on urban geography, for example, cites all of the seminal pieces on central-place theory, such as Ullman's (1941) original paper, and devotes two pages (Mayer, 1954, pp. 152–63) to the three 'models' of intra-urban spatial patterns which had been reviewed a decade previously by Harris and Ullman (1945), but there is not a single reference to any work done by geographers in the context of those models. Thus although there were some precedents in the literature, in general very little work had been done by geographers, prior to the mid-1950s, which followed the dictates of the 'scientific approach'.

Once a new idea gains circulation through the professional journals it is available to be taken up by all. Nevertheless, development of the idea is uaually concentrated in a few places only, where the pioneer teachers encourage their students—in the American case, particularly their graduate students—to conduct research within the new framework. Thus most of the developments in systematic studies in geography during the 1950s can be traced back to a few centres in the United States, and it is the work conducted at those centres which is discussed here. It should be stressed at the outset that the developments to be outlined were largely concerned with method, and that their larger scientific underpinning was stressed very little, although law-seeking was the clear goal. Certainly it was methods which dominated the literature, of both the pros and cons; many of the early contributions of the former groups were in relatively fugitive, departmental publications, presumably because of difficulties in getting such 'new' material accepted by the journals.

The Iowa school
Although Schaefer was at Iowa until his death in 1953, he was not the major influence on the developments which occurred there among the geographers who were, for a number of years, part of the economics department and thus open to the views and approaches of their peers in that more 'mature' social science. The leader of this group of geographers was Harold McCarty, author of a major text on American economic geography (McCarty, 1940). Associated with him were included J. C. Hook, D. S. Knos, H. A. Stafford, and, later, J. B. Lindberg, E. N. Thomas and L. J. King.

The intent of McCarty and his co-workers was to establish the degree of correspondence between two or more geographical patterns, akin to the morphological laws of accordance discussed by Schaefer. (Interestingly, none of their publications refers to Schaefer's paper, although they do refer to the works of, and assistance given by, Gustav Bergmann, a logical positivist of the Vienna School who also strongly influenced Schaefer and read the proofs of his 1953 paper: Davies, 1972, p. 134.) These laws were to be embedded in a theory; thus (McCarty, 1954, p. 96):

> If we are to accept the idea that economic geography is becoming the branch of human knowledge whose function is to account for the location of economic activities on the various portions of the earth's surface, it seems reasonable to expect the discipline to develop a body of theory to facilitate the performance of this task.

Such a theory could be either topically or areally focused, and in its early stages of development would probably be restricted both in its areal coverage and in the topics whose spatial interrelationships it considered.

The purpose of theory is to provide explanations, and McCarty recognized two sorts of explanation. The first is based on a search for the cause of observed locational patterns, but:

> the search for causes can never produce an adequate body of theory for use in economic geography. . . . Variables became so numerous that they were not manageable, and, in consequence, solutions to locational problems were not obtainable (p. 96).

The second, and preferred, type focuses on associations:

> Its proponents take the pragmatic view that if one knew that two phenomena always appear together in space and never appear independently, the needs of geographic science would be satisfied, and there would be scant additional virtue in knowing that the location of one phenomenon caused the location of another (p. 97).

Such laws of association are built up in a series of stages, which begins with a statement of the problem and of the necessary operational definitions, and proceeds through the measurement of the phenomena (with attendant problems of sampling in time and space) to a statement of the findings, in tabular or graphical form. These three descriptive stages precede analysis which seeks out correlations between the distributions of phenomena:

> the nub of the problem of research procedure seems to lie in finding the best techniques for discovering a, b and c in the 'where a, b, c, there x' hypothesis in order to give direction to the analysis. But where shall we search for its components? . . . One source of . . . clues lies in the findings of the systematic sciences. The other source lies in the observations of trained workers in the field or in the library (p. 100).

Thus geography, in its search for morphological laws, is to a considerable extent a consumer of the laws of other disciplines. These laws may be theoretically rather than empirically derived: according to the causal or process approach to explanation:

> Models may be created showing optimal locations for any type of economic activity for which adequate cost data may be obtained. These models may then be used (as hypotheses) for the comparison of hypothetical locations with actual locations. Divergences of pattern may then be noted and the hypothesis altered to allow for them (often by inclusion of factors not ordinarily associated with monetary costs). Ultimately the hypothesis becomes generally applicable and thus takes on the status of a principle (McCarty, 1953, p. 184).

This statement, although not referenced as such, very faithfully reflects the views of logical positivists, and also of Popper, as to how science progresses by the continual modification of its hypotheses, so as better to represent reality.

In their major demonstration of this procedure in operation, McCarty *et al*. (1956) discussed several statistical procedures for measuring spatial association and adopted the now well known technique of multiple regression and correlation which had been used previously, among geographers, by Rose (1936) and Weaver (1943)—both apparently as a product of contacts with agricultural economists. Their empirical context was similarity in the location patterns of certain manufacturing industries in the United States and in Japan: other studies by the group included Hook's (1955) on rural population densities, Knos's (1968) on intra-urban land-value patterns, and King's (1961) on the spacing of urban settlements. Thomas (1960) had used similar procedures in his study of population growth in suburban Chicago, presented as a PhD thesis to Northwestern University, and he extended the methodology with a paper on the use of residuals for regression for both identifying where the laws of association do not apply fully and suggesting further hypotheses for areal associations. (This last paper developed on an earlier one by McCarty (1952) which was not widely circulated.) Later, McCarty (1958) expressed some doubts about the statistical validity of the procedure, but the method that he and his associates pioneered, with its focus on the testing of simple hypotheses derived either from observation or from theoretical deductions, became the model for much research in the ensuing decades.

Wisconsin

The Department of Geography at the University of Wisconsin, Madison, has a long tradition of research with a quantitative bent. Notable among its early products was the PhD thesis of John Weaver on the geography of American barley production, which included a major section (published in his 1943 paper with no supporting methodological argument) using multiple correlation and regression to identify the influences of climatic variables on barley yields. (Weaver later taught at the University of Minnesota, where he developed a widely adopted statistical procedure (Weaver, 1954) for the definition of agricultural regions.) Other work at Madison focused on the measurement of population patterns (e.g. Alexander and Zahorchak, 1943). A combination of these two interests was furthered by a group led by A. H. Robinson, whose main interests have been in cartography, and cartographic correlations were apparently introduced to him by his research supervisor at Ohio State University, Guy Harold Smith (Brown, 1978); Robinson was assisted by R. A. Bryson, of the Department of Meteorology, who was a source of statistical ideas and expertise.

Robinson's concern was to develop statistical methods of map comparison, as indicated by the title of his early paper—'A method for describing quantitatively the correspondence of geographical distributions' (Robinson and Bryson, 1957). As with the work done at Iowa, the lead of Rose and of Weaver was followed with the adoption of correlation and regression procedures. Particular attention was paid to the problems of representing areal data by points (Robinson *et al.*, 1961) and of using correlation methods in the comparison of isarithmic maps (Robinson, 1962). Like McCarty (1958), Robinson was aware of difficulties in applying classical statistical procedures to areal data, and he proposed a procedure to circumvent one of these (Robinson, 1956): Thomas and Anderson (1965) later found this proposal wanting, as it dealt with a special case only and not with the more general problems. Interestingly, though, the main early work on this topic was published by a group of sociologists, under the title of *Statistical Geography* (Duncan, Cuzzort and Duncan, 1961: perhaps even more interestingly, this work was virtually ignored by geographers).

W. L. Garrison and the Washington school
By far the largest volume of work in the spirit of Schaefer's and McCarty's proposals that was published during the 1950s came from the University of Washington, Seattle. The leader of the group of workers there was W. L. Garrison, whose PhD was from Northwestern University, and who, according to Bunge (1966, p. ix), was influenced by Schaefer's paper, although the dates of his earliest publications indicate that he was involved in applying the logical-positivist method to systematic studies in human geography before 1953. Also involved was E. L. Ullman, who moved to Seattle in 1951 (Harris, 1977), and who had already done pioneering research into urban location patterns and transport geography, and the general support to Garrison's efforts was given by the departmental chairman, G. Donald Hudson. A large group of graduate students worked with Garrison and several became leaders in the new methodology during the subsequent decade, including B. J. L. Berry, W. Bunge, M. F. Dacey, A. Getis, D. F. Marble, R. L. Morrill,

J. D. Nystuen, and W. R. Tobler. The group also benefited from a visit to Seattle by the Swedish geographer, Torsten Hägerstrand, who was developing methods of generalizing spatial patterns and processes (see below, p. 85) and from Garrison's contacts with the Business School and the Engineering department at Seattle (Halvorson and Stave, 1978).

Garrison and his co-workers had catholic interests in urban and economic geography. Much of their work was grounded in theory which they gleaned from other disciplines—notably economics—and they directed their efforts towards both testing those theories and applying them to problems of planning. (Another staff member at Seattle, M. E. Marts, had a major interest in regional planning.) In developing testable theoretical statements they displayed a much stronger mathematical base than was the case at Iowa and Madison. They also searched widely for statistical tests which were relevant to their investigation of point and line patterns—the biological sciences provided several of those which they adopted, such as for nearest-neighbour analysis of point patterns (Dacey, 1962) and others, used for grouping and classifying, were derived from psychology (Berry, 1968). Garrison's (1956a, p. 429) view was that 'there is ample evidence that present tools are adequate to our present state of development. No type of problems has been proposed that could not be treated with available tools,' which was in contradiction to an earlier claim by Reynolds (1956), although he was critical of how some standard techniques had been used (Garrison, 1956b). The dominant thrust of the group's work, therefore, involved the derivation from other systematic sciences of relevant theories, mathematical methods and statistical procedures with which to develop morphological laws.

The wealth of the work done at Seattle is best illustrated by a few of their major publications. Garrison contributed an important three-part review article, for example (Garrison, 1959a, 1959b, 1960a), on the state of location theory. The first part comprised a review of six recent books—none of them by geographers—devoted to the question 'What determines the spatial arrangement (structure, pattern, or location) of economic activity?' (Garrison, 1959a, p. 232). Each of the books incorporated locational considerations into traditional economic analysis, and Garrison concluded that this offered valuable economic insights to traditional geographical problems; for geographers to advance their locational analyses and theories they would have to pay much more attention to the work of economists.

Among the Washington group, central-place theory was the dominant location theory on which they worked. This has several independent origins, as Ullman (1941) had indicated (see also Harris, 1977, and Freeman, 1961, p. 201, who notes that findings akin to those in central-place theory were reported by the 1851 Census Commissioners of Great Britain). It was Christaller's (1966) thesis which attracted most attention, however. Working in Germany in the 1930s, Christaller developed ideas regarding the ideal distribution of settlements of different sizes acting as the marketing centres of functional regions, within the constraints of assumptions relating to the physical environment and the goals of both entrepreneurs and customers: a translation became available in the late 1950s, and was published in 1966. Studies of functional regions were not novel, of course (see above, p. 35), and geographers had already attempted to test Christaller's notions regarding a hierarchical organization of settlements distributed on a hexagonal lattice (e.g. Brush, 1953). Dacey was concerned to make the analysis of these spatial hierarchies more rigorous (Dacey, 1962) and Berry's research focused on the central-place aspects of the settlement pattern north of Seattle and on the retail centres in the city of Spokane (Berry and Garrison, 1958a, 1958b; Berry, 1959a).

The second part of Garrison's (1959b) review article dealt with possible geographical applications of the mathematical procedures of linear programming, which produce the optimal solutions to problems of resources allocation in constrained situations. In this, he illustrated how the procedures of neoclassical economic analysis could be adapted in order to investigate ideal solutions to the problems of where to locate economic activities and how to organize flows of goods. Six such problems which could be treated by linear programming were identified:

1 the transportation problem, which takes a set of points, some with a given supply of a good and some with a given demand for it, plus costs of movement, and determines the most efficient flow pattern of the good from supply to demand points which minimizes the expenditure on transport;

2 the spatial price-equilibrium problem, which takes the same information as the transportation problem, but is used to determine prices as well as flows;

3 the warehouse-location problem, which determines the best location for a set of supply points, given a geography of demand;

4 the industrial-location problem, which determines the optimum location for factories from knowledge of the sources of their raw materials and the destinations for their products;

5 the interdependencies problem, which locates linked plants so as to maximize their joint profits; and

6 the boundary-drawing problem, which determines the most efficient set of boundaries (i.e. that which minimizes total expenditure on transport) for, for example, school catchment areas. The purpose of these analyses, if they are being used to investigate actual patterns and not as the bases for future plans, was, as Lösch (1954) put it, to see whether reality was rational, whether decision-makers had acted in ways that would produce the most efficient solutions, where efficiency was defined as cost-minimization, particularly transport-cost minimization. Investigations by the Washington group in this context included studies of interregional trade (Morrill and Garrison, 1960) and the optimal

location, by regions, of agricultural activities in the United States (Garrison and Marble, 1957).

In the final part of his review article, Garrison (1960a) dealt with four further books on locational analysis, which were empirical in orientation and shared a common interest in the agglomeration economies reaped by industrial clusters. Several topics and techniques were discussed, such as the use of input-output matrices to represent industrial systems, and Garrison concluded by stressing the need for geographers to investigate location patterns as systems of interrelated activities, a topic which is taken up again in the next chapter.

The empirical work with a planning orientation undertaken by the Seattle group is illustrated by one of the outputs of Garrison's large study of the impact of highway developments on land use and other spatial patterns (Garrison *et al.*, 1959). This book includes four studies: Berry's on the spatial pattern of central places within urban areas; Marble's on the residential pattern of the city (as indexed by property values) and relationships between household characteristics—including location—and their movement patterns; Nystuen's on movements by customers to central places; and Morrill's on the locations of physicians' offices, both actual and the most efficient. In addition, Garrison himself worked on the accessibility impacts of highway improvements, and devised indices of accessibility based on graph theory (Garrison, 1960b: the work was continued by Kansky, 1963, after Garrison and Berry moved to work in the Chicago area). He also used the simulation procedures that had been developed by Hägerstrand (1968) to investigate urban growth processes (Garrison, 1962), a topic taken much further by Morrill (1965).

Somewhat separate from the work of the others in the group, although aligned with their general purpose, was Bunge's thesis, *Theoretical Geography* (1962, reprinted in enlarged form, 1966). This too displays a catholic view of geography, together with an acknowledged debt to Schaefer (Bunge worked at Iowa for a short period). It is an extremely difficult book to summarize, but the basic theme is very clear: geography is the science of spatial relations and interrelations; geometry is the mathematics of space; hence geometry is the language of geography. Thus the early chapters are concerned with establishing geography's scientific credentials, in a debate with Hartshorne's published statements, especially those concerning uniqueness and predictability. As Lewis (1965) and others also have argued, Bunge claimed that Hartshorne confused uniqueness and singularity: he opposed Hartshorne's claim that geography cannot formulate laws because of its paucity of cases by arguing for even more general laws, and countered the argument that geographical phenomena are not predictable with the claim that science 'does not strive for complete accuracy but compromises its accuracy for generality' (p. 12).

Having established geography's scientific credibility to his satisfaction, Bunge then investigated its language. An intriguing discussion of cartography led him to conclude that descriptive mathematics is preferable to cartography as a more precise language. The remainder of the book looked at aspects of the substantive content of the science of geography, beginning with 'a general theory of movement' and then a chapter on central-place theory:

> If it were not for the existence of central-place theory, it would not be possible to be so emphatic about the existence of a theoretical geography . . . central-place theory is geography's finest intellectual product (p. 133).

Problems of testing the theory were then discussed, showing, as did Getis (1963), the need for map transformations, and in the final chapter of the first edition the links between geography and geometry were made clear:

> Now that the science of space is maturing so rapidly, the mathematics of space—geometry—should be utilized with an efficiency never achieved by other sciences (p. 201).

The richness of the work done by this group during the mid and late 1950s continued in various locations after it broke up (only Ullman remained, although Morrill later joined the staff at Seattle). Most prolific and seminal has been Berry, not only in his original field of central-place theory (Berry, 1967) but also over a very wide range of other topics in economic and social geography; he has become one of the most cited geographers of all time. Berry's work has always had a very strong empirical and utilitarian base, but Dacey has continued to work on the mathematical representation of spatial, especially point, patterns (e.g. Dacey, 1973). In total, the work of this group of scholars influenced the research and teaching of a whole generation of human geographers, throughout the world, and has undoubtedly been the most significant in the development of the first post-regional paradigm.

The social physics school
The work of this group was initiated and developed independently from that of any of the other three, and was not apparently influenced by Schaefer—its early publications preceded Schaefer's paper by more than a decade. The leader was J. Q. Stewart, an astronomer at Princeton University, who traced the origins of social physics in the work of a number of natural scientists who applied their methods to social data (Stewart, 1950). His own work apparently began when he noted certain regularities in various aspects of population distributions, regularities which were akin to the laws of physics, such as a tendency for the number of students attending a particular university to decline with increasing distance of their home addresses from its campus. From these observations he developed his ideas on social physics, which he defined (Stewart, 1956, p. 245):

. . . that the dimensions of society are analogous to the physical dimensions and include numbers of people, distance, and time. Social physics deals with observations, processes and relations in these terms. The distinction between it and mathematical statistics is no more difficult to draw than for certain other phases of physics. The distinction between social physics and sociology is the avoidance of subjective descriptions in the former.

and he established a laboratory at Princeton to investigate the wide range of regularities which could be analysed in this context.

Stewart's ideas were introduced to geographers by a paper in the *Geographical Review* (Stewart, 1947). Four empirical rules were adduced there: the first, the rank-size rule for cities, showed that in the United States the population of a city multiplied by its rank (from 1 for the largest to n for the smallest), and standardized by a constant, equalled the population of the largest city, New York; the second indicated that at various dates the number of cities in the country with populations exceeding 2500 was very closely related to the proportion of the population living in such places; the third showed that the distribution of a population could be described by the population potential at a series of points, in the same manner as the potential in a magnetic field is described in Newtonian physics; and the fourth illustrated a close relationship between this population potential and the density of rural population in the United States. From these regularities, Stewart claimed that

> There is no longer any excuse for anyone to ignore the fact that human beings, on the average and at least in certain circumstances, obey mathematical rules resembling in a general way some of the primitive 'laws' of physics (p. 485).

No reasons were given why this should be so (Curry, 1967b, p. 285, called this a 'deliberate shunning of plausible argument'): the rules were presented as empirical regularities which had some similarity to the basic laws of physics. Causal hypotheses were not even postulated, let alone tested.

One of Stewart's collaborators was William Warntz, a graduate of the University of Pennsylvania who was later employed by the American Geographical Society as a research associate, working on what he termed 'the investigation of distance as one of the basic dimensions of society' (Warntz, 1959b, p. 449). The wide range of empirical regularities which they observed (see Stewart and Warntz, 1958, 1959) was used to develop their concept of macrogeography (Warntz, 1959b, 1959c). Geographical work, Warntz claimed, was dominated by micro-studies:

> The tendency of American geographers to be preoccupied with the unique, the exceptional, the immediate, the microscopic, the demonstrably utilitarian, and often the obvious is at once a strength and a weakness (Warntz, 1959b, p. 447)

but

the assembly of more and more area studies involving an increase in the quantity of detail does not mean *per se* a shift from the microscopic to the macroscopic (p. 449).

Thus geographers were in danger of being unable to perceive general patterns within their welter of local detail. To counter this, Stewart and Warntz suggested the search for 'regularities in the aggregate'. Stewart's concept of population potential was used to describe general distributions, and was shown to be related to a large number of other patterns in the economic and social geography of the United States. These findings, it was realized, were only empirical regularities, but they could be used as the basis for theory development (Stewart and Warntz, 1958, p. 172), for geography's need was theory which 'has as its aim the establishment and coordination of areal relations among observed phenomena. General laws are sought that will serve to unify the individual, apparently unique, isolated facts so laboriously collected' (Warntz, 1959b, p. 58). The approach to theory, then, was inductive (see Figure 3.1) rather than deductive. To Warntz, as to Garrison, Ullman, and others, the ideas of Christaller were seminal (see Bunge, 1968).

These macroscopic measures, particularly that of population potential, were used in a variety of contexts, as in Warntz's (1959a) *Towards a Geography of Price*, which established strong relationships between the prices of agricultural commodities in the USA and measures of supply and demand potential; Harris (1954b) and Pred (1965a) also used the potential measure in their studies of industrial-location patterns. The 'macrogeographers' also did much work on various distance-decay functions (see Chapter 4) and developed the early work of a Russian group on centrographic measures (Sviatlovsky and Eels, 1937; Neft, 1966).

These lines of work contrasted markedly with that of the other three groups reviewed here, in a variety of ways. First was the topic of scale; Stewart and Warntz perhaps conformed more than any others to Bunge's call for a scientific approach which aimed at a high level of generality. Second there was the nature of the approach to theory, for macrogeography was inductive in its search for regularity rather than testing deductive hypotheses. Finally, the analogies sought for human geography were in a natural science—physics—and not in the other social sciences.

Summary

The developments outlined in this section marked the beginning of major changes in the field of human geography, changes which were rapidly taken up by others, within and beyond the United States. Although the focus was on theory and measurement, and the development of 'geographical laws', in line with the general ethos of academia in the immediate post-war decades, to some extent the work did not deviate too far from Hartshorne's expanded definition of the nature of geography (see above, p. 46). The main difference between the new work, with its focus in the systematic studies, and the regional paradigm was the greater faith of geographers in their ability to produce laws, to work within the canons of accepted scientific method, and to move out of their self-imposed academic isolation (Ackerman, 1945, 1963).

The scientific method in geography

Whether or not the changes just described initiated a 'Kuhnian revolution' in human geography is a topic left to the last chapter of this book. It is clear, however, that they involved at least a major reorientation in the nature of geographical research. Perhaps surprisingly, this reorientation was focused on no programmatic statement. There is no published paper or book which provided a detailed outline of how research should be conducted in this new framework: Schaefer's paper said nothing about how geographical laws were to be stated and derived, and although McCarty and his co-workers discussed methods in both general (McCarty, 1954) and specific (McCarty *et al.*, 1956) contexts, they did not provide an overall programme. As Gregory (1978a, p. 47) puts it, 'geography has (with some notable exceptions) paid scant attention to its epistemological foundations'.

One piece was not central to the initial efforts, because it post-dated many of them, but it was widely quoted in the 1960s, as the new ideas spread. Ackerman's (1958) essay was entitled *Geography as a Fundamental Research Discipline* and was an analysis of research organization. He indicated that the ultimate goal was integration to provide a full comprehension of reality, and that with regard to current developments, 'If any one theme may be used to characterize this period, that theme would be one of illuminating covariant relations among earth features' (p. 7). As a science, even one which is eventually an idiographic science since it deals with unique places, geography, according to Ackerman, needed to strive for 'an increasingly nomothetic component'. Its fundamental research

> need not necessarily be law-giving. . . . Much fundamental research in geography has not been law-giving in the strict sense but it has been concerned with a high level of generalization, and it has given meaning to other research efforts which succeeded it. In this sense it has a block-building characteristic (p. 17).

Such fundamental research 'is likely to rest on quantification . . . accurate study depends on quantification' (p. 30) and will 'furnish a theoretical framework with capacity to illuminate actually observed distributional patterns and space relations' (p. 28).

Ackerman's essay was a clarion call for the development of theory, for the application of quantitative methods, and for a focus on laws and generalizations to form the building-blocks for further nomothetic research. But there was no detailed discussion of how such research should be undertaken: it was a clarion call without reference to detail. Seven years later, the report of a National Academy of Sciences–National

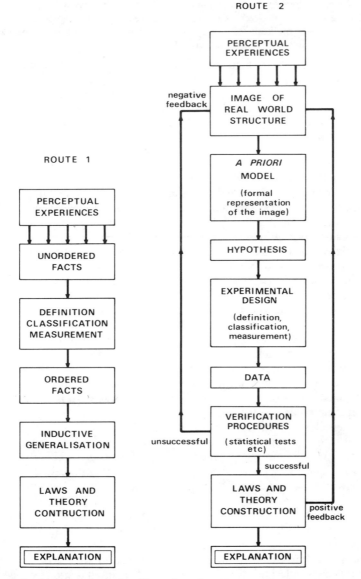

Figure 3.1   Two routes to scientific explanation. Source: Harvey (1969a, p. 34).

Research Council (1965) committee on *The Science of Geography* discussed 'Geography's problem and method' with the statement that

> Geographers believe that correlations of spatial distributions, considered both statistically and dynamically, may be the most ready keys to understanding existing or developing life systems, social systems, or environmental changes. In the past . . . progress was gradual, however, because geographers were few, rigorous methods for analysing multivariate problems and systems concepts were developed only recently (p. 9).

Again, a general statement on orientation of research, but no detail on how research was to be conducted. And yet, in a paper published in 1963, Burton claimed that an intellectual revolution—the quantitative and theoretical revolution—had occurred in geography: 'The revolution is over, in that once-revolutionary ideas are now conventional' (p. 156). Something had become conventional, but nobody had written a full formulation for the discipline of what that something was!

The scientific method
There is no suggestion here that the groups of researchers who proposed changes in the nature of human geography had no clear rationale for their work; indeed, those involved were undoubtedly very clear as to both means and ends (though this may not be the case with some of their disciples), but they did not discuss these in detail in print. Nor, if their citations in the published works are any lead, did they research deeply into the philosophy that they were adopting: the exceptions to this are the references to Bergmann in the Iowa group's papers and in Bunge's thesis. Texts on statistical and mathematical procedures were widely quoted, but the first major work on the philosophy of the 'new geography' was not published until 1969, in a book which received wide acclaim (Harvey, 1969a: a parallel, but briefer statement is Moss, 1970).

There are two routes to explanation, according to Harvey (Figure 3.1). The first, occasionally known as the 'Baconian' or inductive route, derives its generalizations out of observations: a pattern is observed and an explanation developed from and for it. This involves a dangerous form of generalizing from the particular case, however, because, as Moss (1970) argues, acceptance of the interpretations depends too much on the charisma of the scholar involved: so the preferred method is the second route in Figure 3.1. This also begins with an observer perceiving patterns in the world; he then formulates experiments, or some other kind of test, to prove the veracity of the explanations which he produced for those patterns. Only when his ideas have been tested successfully against data other than those from which they were derived can a generalization be produced.

Scientific knowledge, obtained via the second route, is 'a kind of controlled speculation' (Harvey, 1969a, p. 35), and it is such a procedure that an increasing number of human geographers sought to apply during the 1950s. The method, known as logical positivism, was developed by a group of philosophers working in Vienna during the 1920s and 1930s (Guelke, 1978). It is based on a conception of an objective world in which there is order waiting to be discovered. Because that order—the spatial patterns of variation and covariation in the case of geography—exists, it cannot be contaminated by the observer. A neutral observer, on the basis of either his observations or his reading of the research of others, will derive a hypothesis (a speculative law) about some aspect of reality and then test that hypothesis: verification of his hypothesis translates the speculative law into an accepted one.

A key concept in this philosophy is that laws must be proven through objective procedures, and not accepted simply because they seem plausible: as Bunge (1962, p. 3) puts it, 'the plausibility or intuitive reality of a theory is *not* a valid basis for judging a theory'. A valid law must predict certain patterns in the world, so that having developed an idea about those patterns, the researcher must formulate them into a testable *hypothesis*—'a proposition whose truth or falsity is capable of being asserted' (Harvey, 1969a, p. 100). An experiment is then designed to test the hypothesis, data are collected, and the validity of the predictions evaluated.

If the results of the test do not match the predictions, then either the observations on which the hypothesis was based or the deductions from the works of others are thrown into doubt. There is thus negative feedback (Figure 3.1) and the image of the world has to be revised, creating a new hypothesis. (This in fact is the Popperian view that any hypothesis is found wanting by a single falsification. Harvey (1969a, p. 39) gives only eight lines to this view, however, preferring the more general one that only 'severe failure'—which he does not define—discredits an hypothesis totally: see, however, Moss, 1977; Bird, 1975.) Clearly, if the researcher is a good observer and a logical thinker, such falsification of his hypotheses should be rare. If the test is successful, on the other hand, then the speculation of the hypothesis becomes an acceptable generalization. One successful test will not turn it into a *law*: replication on other data sets will be needed, since a law is supposed to be a universal.

According to Harvey (1969a, p. 105):

> A scientific law may be interpreted as a generalization which is empirically universally true, and one which is also an integral part of a theoretical system in which we have supreme confidence. Such a rigid interpretation would probably mean that scientific laws would be non-existent in all of the sciences. Scientists therefore relax their criteria to some degree in their practical application of the term.

After sufficient (undefined) successful tests, therefore, a hypothesis may be accorded law-like status, and is fed into a body of *theory*, which comprises a series of related laws. There are two types of statement within a full theory: the *axioms*, or givens, which are statements taken to be true, such as laws; and the deductions, or *theorems*, from those initial conditions, which are derived consequences from agreed facts—the next round of hypotheses. There is a positive feedback from the theory stage to the world view, therefore (Figure 3.1), so that the whole scientific enterprise, aimed at total explanation as Ackerman argued, is a circular procedure whereby the successes of one set of experiments become the building blocks for thinking about the next.

One stage in Figure 3.1 so far ignored is the *model*, a widely used term which has been given a variety of meanings (Chorley, 1964). There are two basic functions of a model: as a *representation of the real world*, such as a scale model, a map, a series of equations, or some other analogue; and an *ideal type*, a representation of the world under certain constrained conditions. Both are used in the positivist method to operationalize a theory, as a guide to the derivation of testable hypotheses.

Quantification has a central function in this scientific method. Mathematics are particularly useful in the writing of models, as in the linear-programming procedures adopted by Garrison. Relatively few geographers have strong backgrounds in mathematics, however (this was especially true in the 1950s), and so little work was done which involved representation of the real world as sets of equations. Instead the central role was given to statistics, used in hypothesis-testing. Two types of statistics are available: *descriptive statistics* can be used to represent a pattern or relationship; *inductive statistics* are used to make generalizations about a carefully defined population from a properly selected random sample of that population. Many geographical researchers confused the two. Inductive statistics use significance tests to show whether what has been observed in the sample probably also occurs in the parent population, so that if the data analysed are not a sample, such tests are irrelevant (some disagreement has been expressed over 'what is a sample?': see Meyer, 1972 and Court, 1972). Many geographers have used inductive methods in a descriptive manner, however, using the significance tests (in error) as measures of the validity of their findings.

The main attraction of statistics to many of the early adherents of the 'new geography' was their precision and lack of ambiguity—compared to the English language—in description. This was expressed by Cole (1969) who annotated a quotation from a well known text (Stamp and Beaver, 1947, pp. 164–5): in this the text is Stamp and Beaver's and the annotations in parentheses, to show the ambiguities, are Cole's:

> The present distribution of wheat cultivation in the British Isles (space) raises the conception of two different types of limit. Broadly speaking (vague), it may be said that the possible (vague) limits (limit) of cultivation of any crop are determined by geographical (vague), primarily by climatic, conditions. The limits so determined (how?) may be described (definition) as the ultimate (vague) or the geographical (vague) limits . . . (Cole, 1969, p. 160).

The full quotation, according to Cole (only part is reproduced here), is so full of ambiguities that it could refer to about one million million possible combinations of some forty counties:

> the correlations suggested are so tentative and imprecise that they leave the reader still wondering why wheat is grown where it is. The application of a standard correlation procedure . . . in itself would give a more precise appreciation of the relationship (p. 162).

Similar views became widely held during the 1950s and 1960s, and quantification became the *sine qua non* of training in the new methods (LaValle, McConnell, and Brown, 1967).

The scientific method increasingly adopted by geographers is a procedure for testing ideas, therefore, but a highly formalized one, about which there has been a great deal of debate (Harvey, 1969a). Although many aspects of the method were used by geographers, their citations indicate relatively little research in depth into the full philosophy of logical positivism.

Reactions to the scientific method
Despite (or perhaps because of) the lack of a clear programmatic statement of the 'new Theology' (Stamp, 1966, p. 18) until Harvey's (1969a) book, reactions to the developments were many and varied. (James, 1965, p. 35, called the debate 'continued, bitter, and uncompromising warfare'.) Two related issues were the main foci of contention: whether quantification was sensible in geographical research, and whether lawmaking was possible. As Taylor (1976) points out, to some extent the debate was inter-generational, of the type discussed in Chapter 1 (p. 21): to some of the 'old guard' what was being proposed was just not geography and should be banished to some other corner of academia.

The quantification issue was the less important, and few spoke out against it in its entirety, although its extent was criticized. Thus Spate (1960a, p. 387) recognized that quantification is 'an essential element' and

> This is, like it or not, the Quantified Age. The stance of King Canute is not very helpful or realistic; better to ride the waves, if one has sufficient finesse, than to strike attitudes of humanistic defiance and end, in Toynbee's phrase, in the dustbin of history (p. 391).

He found three dangers in the development, however. The first was a confusion of ends and means. Quantifiers wanted to quantify everything (after Lord Kelvin—'when you cannot express it in numbers, your knowledge is of a meagre and unsatisfactory kind': Spate, 1960b), but some things, like the positions of Madrid and Barcelona in Spanish thought, cannot be treated in that way. Secondly, there was the dogged analysis of trivia, producing platitudinous findings, a fault which Spate recognized is part of all academia, and especially its revolutions:

'Quantified or not, the trivial we will always have with us' (Spate, 1960a, p. 389), and the problem is usually the extreme positions taken up by the protagonists—Robinson's (1961) perks (the hyperquantifiers) and pokes (the hypoquantifiers). Finally, there was the vaunting ambition of the quantifiers, the belief that solution of the world's problems lay just around the corner.

Spate was not the only critic, and he was more generous than many. Burton (1963) identified five types of critic:

1   those who felt geography was being led in the wrong direction;
2   those who felt geographers should stick to their proven tool—the map;
3   those who felt that quantification was suitable for certain tasks only;
4   those who felt that means were being elevated over ends, and there was too much research on methods for methods' sake; and
5   those who objected not to quantification but to the quantifiers' attitudes. He believed, however, that quantification had been proven to be more than a fad or fashion and that geography would develop out of a stage of testing relatively trivial hypotheses with its new tools so that 'The development of theoretical, model-building geography is likely to be the major consequence of the quantitative revolution' (p. 156).

More critical to many geographers than quantification was this issue of theory, and in particular the question of the role of laws in geography. For some, this continued the debate over environmental determinism, which was still active in Britain (Clark, 1950; Martin, 1951; Montefiore and Williams, 1955; Jones, 1956). Jones, for example, carried this debate over to the topic of scientific determinism, and its implications with regard to human free will. Martin (1951) had argued that possibilism is 'not merely wrong but is mischievous' (p. 6) because all human actions are determined in some way, so that in human geography:

> Unless we can assume the existence of laws or necessary conditions similar in stringency to those of physical science, there can be no human geography nor social sciences worth the name, but only a series of unexplainable statements of bare events . . . such laws cannot differ, except in respect of . . . far greater complication, from those of physical science (pp. 9–10).

Jones (1956) pointed out the impossibility of discovering universal laws about human behaviour and indicated the existence of two types of law of physics: the determinate laws of classical physics, which applied macroscopically; and the probabilistic quantum laws which refer to the behaviour of individual particles. Use of the latter would allow for the exercise of free will within prescribed constraints, and would at least allow answers to be offered to the question 'how?' if not to 'why?'. But the question of causality clearly worried many, as indicated by Lewis's (1965) counterargument that 'it is erroneously assumed that causes compel their effects in some way in which effects do not compel their causes' (p. 26).

Golledge and Amedeo (1968) addressed this same problem, by pointing out that critics of law-seeking in human geography applied a definition of a law as a universal postulate which brooked of no exception. They indicated that science recognizes several types of law, and also that the veracity of a law-like statement can never be finally proven, since it cannot be tested against all instances, at all times and in all places. They indicated four types of law which have relevance for human geographers. *Cross-sectional laws* describe functional relationships (as between two maps) but show no causal connection, although they may suggest one. *Equilibrium laws* state what will be observed if certain criteria are met, whereas *dynamic laws* incorporate notions of change, with the alteration in one variable being followed by (perhaps causing) an alteration in another. Dynamic laws may be *historical*, showing that B would have been preceded by A and followed by C, or *developmental*, in which B would be followed by C, D, E etc. Finally, there are *statistical laws* which are probability statements of B happening, given that A exists: all laws of the other three categories may be either deterministic or statistical, with the latter almost certainly the case with phenomena studied by geographers.

None of the papers just discussed was part of an ongoing debate on quantification and theory-building; they appear to have been reactions to attitudes rather than to published critiques (in Britain there were none for several years: Taylor, 1976). There was one debate, however, in the American literature. It was initiated by Lukermann (1958), who was reacting to the views of Warntz on macrogeography (see above, p. 59) and to a paper by Ballabon (1957). The latter had claimed that economic geography was lacking general principles and was 'short on theory and long on facts' (p. 218). McCarty had shown the way on how to conduct research, but stress was placed on the need to study location theory being developed by economists as a source for hypotheses. Lukermann's response was that the main problem in the proposals of Ballabon and Warntz lay in the assumptions behind their hypotheses (Warntz's analogies with physics and Ballabon's with economics) which did not conform to his view of geography as an empirical science. Statistical regularities and isomorphisms with other subject matter do not provide explanations, so that hypotheses derived from such models test only the models themselves (see also Moss, 1970): 'the hypotheses to be tested are neither statistically nor rationally derived; that is, they are derived neither from empirical observation nor from deductions of previous knowledge in the social, economic or geographic fields' (p. 9). Research in economic geography, Lukermann argued, should start with the recording of data on maps.

Lukermann was taken up by Berry (1959b) who argued that models, for all their simplifications and unreal assumptions, can offer insights towards understanding the real world: 'A theory or model, when tested and validated, provides a miniature of reality and therefore a key to many descriptions. There is a single master-key instead of the loaded key ring' (p. 12). Lukermann (1960a) was not convinced that models based on assumptions of perfect knowledge and competition, for example, could help in understanding if they were not empirically derived: 'the crucial problem is the construction of hypotheses from the empirical realities of economic geography . . . more light is shed and less truth is sophisticated through inventory than through hunches' (p. 2). King (1960) then entered the debate, pointing out that all laws are really only hypotheses, and that deviations of observed from expected in their testing indicates where the assumptions are invalid. Lukermann responded three times. In the first paper, he showed the lack of consensus in 'explanations' of the geography of cement production in the United States (Lukermann, 1960b) because economic analyses ignored 'Historical inertia, geographical momentum, and the human condition' (p. 5). His second paper (1961), in response to King, presaged some of the arguments developed later by Sack, who worked with him at Minnesota (see below, p. 96), and pointed out that much of the theory being introduced to economic geography (such as Lösch's) was not based on providing understanding of, and explanation for, reality. Finally, he presented a longer paper (Lukermann, 1965) which discussed several aspects of the debate, concluding with the statement that

> Thus, we see scientific explanation as far removed from the context within which the macroscopic geographers would have us put it—the end product of geographic research. Science does not explain reality, it explains the consequences of its hypotheses (p. 194),

and a further call for explanations in geography to be based on observations of reality and not the import of analogies which cannot offer explanations, but only unreal assumptions.

The clear difference of opinion between Lukermann and his antagonists over the way in which geographers should seek explanations (which was not about the scientific method itself, but about the inputs to the images of the real-world structure—Figure 3.1) suggests the sort of generation gap discussed in Chapter 1. It is doubtful whether papers such as those of Jones, of Lewis, and of Golledge and Amedeo quieted the fears of those unconvinced by the arguments of the 'quantifiers', any more than Berry and King convinced Lukermann. But the differences soon became a non-issue, at least in the published papers resulting from the research activities of geographers in many topical specialisms. As Burton claimed, by the mid-1960s the changes seem to have been widely accepted, and the regional paradigm had certainly been ousted from its prime position in the publications of human geographers. Increasingly, quantitative and theoretical material using the scientific method, in part at least, came to dominate not only the more obvious journals, such as

Economic Geography and *Geographical Analysis* (a 'journal of theoretical geography' founded in 1969), but also the prestigious general journals, notably the *Annals of the Association of American Geographers*. (The *Geographical Review* was an early partial 'convert' through the American Geographical Society's sponsorship of the macrogeographers although Berry states that it rejected his early papers with Garrison as 'not geography': Halvorson and Stave, 1978.) And by the 1970s, textbooks were being published which began with discussions of scientific method and quantification before proceeding to the substantive content of the 'empirical science' (Abler, Adams and Gould, 1971; Amedeo and Golledge, 1975).

Spread of the scientific method

Within human geography in the United States, the initial development of systematic studies using the positivist scientific method was very largely focused on economic geography and the associated economic aspects of urban geography. This undoubtedly reflects the relative sophistication of economics within the social sciences, providing a model for geographers to copy, not only to advance their discipline but also to promote its cause in the search for utility to the world of business and government. The long tradition of empirical work in human geography meant, however, that with few exceptions research in the systematic areas mainly comprised the statistical testing of relatively simple hypotheses, with little mathematical modelling or writing of formal theory.

Contemporaneous with these developments in human geography, and an important stimulus for them, was the emergence of a new discipline in the United States—regional science. This was very much the product of one iconoclast scholar—Walter Isard—an economist who built spatial components into his models, in part to provide a stronger theoretical basis for urban and regional planning than had existed previously. In general terms, regional science is economics with a spatial emphasis, as illustrated by Isard's (1956a, 1960) two early texts, and many members of the Regional Science Association are practising geographers. To some, regional science and economic geography are hard to distinguish: the former can be separately characterized by its greater focus on mathematical modelling and economic theorizing, however, whereas geographical work has remained more empirical and less dependent on formal languages. (Initially Isard, 1956b, saw geographers as doing the empirical tests of the regional scientists' models.) Over time, the interests of regional scientists have broadened (Isard, 1975), but the strong theoretical base has remained.

The emphasis on statistical methods in so much of the new work in American human geography led to its partial rapprochement with physical geography. (One of the leading 'quantitative geographers' of the

1960s, Leslie Curry, was trained as a climatologist.) More physical papers were published in the leading journals, more physical geographers were appointed to university departments, and there was a common interest in the training of graduate students (LaValle *et al.*, 1967). This common interest in procedures was illustrated by the papers given at a conference held in 1960 on Quantitative Geography, which led to the publication of two volumes (Garrison and Marble, 1967a, 1967b) on the development of methods, one for human geography and the other for physical geography. In the former, for example, Berry introduced the family of factor-analysis methods as a way of collapsing and ordering large data matrices; Dacey investigated line patterns and Beckmann the optimal location of routes; Robinson continued his work on the statistical comparison of maps; Mayfield and Thomas extended the analysis of central-place patterns; Marble, Morrill, and Nystuen looked at patterns of movement; and Warntz continued the work on macrogeography. Other conferences and summer schools to train geographers in quantitative techniques were held at this time (on their impact, see Gould, 1969), and American geographers were to the forefront in the launching of an International Geographical Union Commission on Quantitative Methods which held several international conferences and published volumes of papers extending the statistical procedures available to geographers.

Expansion within American geography

The launching of Burton's 'quantitative and theoretical revolutions' took place in a few topical specialisms within American human geography, so one of the first tasks for the established 'revolutionaries' was to spread their 'new Theology' wider through the discipline, by convincing others of the benefits which quantification and the scientific method could bring to their specialisms. One of their major pieces of advocacy was the NAS/NRC report (1965) on *The Science of Geography* which was prepared in order to chart research priorities within the discipline. The case was presented for more 'theoretical-deductive' work to balance the earlier emphasis on 'empirical-inductive analysis', the detailed argument being based on four premises:

(a) Scientific progress and social progress are closely correlated, if not equated. (b) Full understanding of the world-wide system comprising man and his natural environment is one of the four or five great overriding problems in all science. (c) The social need for knowledge of space relations of man and natural environment rises, not declines, as the world becomes more settled and more complex, and may reach a crisis stage in the near future. Last, (d) progress in any branch of science concerns all branches, because science as a whole is epigenetic.

The social need for knowledge of space relations means an imminent practical need. As the population density rises and the land-use intensity increases, the need for efficient management of space will become even more urgent (p. 10).

And since, to the members of the committee (E. A. Ackerman, B. J. L. Berry, R. A. Bryson, S. B. Cohen, E. J. Taaffe, W. L. Thomas Jr., and M. G. Wolman) geography 'involves the study of spatial distributions on the earth's surface' (p. 8) then it followed that 'Geographic studies will be irreplaceable components of the scientific support for efficient space management' (p. 10). The scientific method was being sold to geographers and geography was being sold to the scientific establishment.

The committee chose four problem areas within geography to illustrate the subject's potentials as a 'useful science'. The first of these was physical geography. The second was cultural geography, which studies 'differences from place to place in the ways of life of human communities and their creation of man-made or modified features' (p. 23). A major focus in this area, it was noted, was on landscape development and the diffusion over space and time of specific cultural features: 'applying modern techniques to studying the nature and rate of diffusion of key cultural elements and establishing the evolving spatial patterns of culture complexes' (p. 24) was seen as an avenue for development. The third problem area was political geography, and the committee proposed work on boundaries and resource management. Finally, the committee recognized location-theory studies, an amalgam of work in economic, urban and transport geography in which the 'dialogue' between the empirical and the theoretical had gone furthest 'revealing the potential power of a balanced approach when applied to other geographical problem areas' (p. 44). Location theory indicated the nature of the normal science of the new paradigm, with work on spatial patterns, the links and flows between places in such patterns, the dynamics of the patterns, and the preparation of alternative patterns through model-building exercises which identify efficient solutions.

In the development of the science of geography,

> A major opportunity seen by workers in the location-theory problem area is that of integrating their work more closely with other geographers as they begin to deal with spatial systems of political, cultural, and physical phenomena. . . . This could be achieved . . . by the accelerated diffusion of techniques and concepts to other geographers, and communication on the definition of research problems. The result would be to hasten the confrontation of empirical-inductive studies by theoretical-deductive approaches throughout geography. . . . Testing the theory in a variety of empirical contexts should aid in the overall development and refinement of viable theories. It should also serve to connect geographic progress to local problems more rapidly and more effectively (pp. 50–1).

The deductive-theoretical scientific methodology was central to their blueprint for the advancement of geographical research, therefore. All geographers would have a role to play in this movement forward, for:

> Geographers have one other asset that should be capitalized on. Those who have been interested in the study of a specific part of the earth (regional

geography) develop competences for interpreting the physical-cultural complexes of the regions that they study. Students of the way a particular part of
the earth has evolved (historical geography) have other competences for
interpreting the historical development and modification of a region. These
two groups have students that are particularly qualified to undertake the field
observation and field study of problems recognized in a more systematic way
and to conduct field tests of generalizations arrived at through systematic
study. . . .

The regional or historical geography specialist who has mastered the technique of field observation and historical study thoroughly . . . can make
himself indispensable if he understands the direction in which the generalizing clusters are headed and relates his work closely to their growing edges (p.
61).

A clear division of labour was being suggested, comprising theoretical-
deductive 'thinkers' and empirical-inductive 'workers', a division which
was apparently unequal in status and was resented by some as such
(James, 1965; Thoman, 1965).

One of the systematic areas of geography that was colonized early by
the new methods was that part of urban geography which dealt with the
internal spatial structure of cities. Until the 1960s, little work had been
done on this topic except with regard to commerical land uses (particularly the Central Business District and the relationship of the pattern of
suburban shopping centres to the postulates of central-place theory):
almost no attention was paid to the human content of residential areas,
perhaps because geography was seen to be the science of places, not of
men. Recognition that 'people live in cities' (Johnston, 1969) led to a
growth of interest in residential areas, which gained much stimulus from
the work of the Chicago urban-ecology school of sociologists (some of
their works had been introduced to geographers earlier—Harris and
Ullman, 1945: Dickinson, 1947—but with little success). The methods
of social-area analysis gained in popularity, using the techniques being
applied elsewhere (Berry, 1964a): a new form of urban geography was
initiated which adopted its norms from the functionalist school of sociology. Society is made up of socio-economic classes, the nature and com-
position of which are widely accepted, and these classes come to consensual agreements about the allocation of land among competing groups
(Johnston, 1971). This type of urban geography in effect became a
separate systematic branch of the discipline; few people did research in it
as well as in other aspects of urban geography.

The argument being advanced by those who sought to spread the new
methodology more widely was focused on a common set of procedures to
tackle geographical problems. This was described by Berry (1964b), who
argued that the geographer's viewpoint emphasizes space, with regard to
distributions, integration, interactions, organization, and processes. All
of its data can be categorized by a single matrix (Figure 3.2) in which
places form the rows and characteristics the columns: each cell defines a

geographic fact. Berry then suggested that five different types of geographical study can be recognized by focusing on different elements of this
matrix: study of a single row (a place) or column (a characteristic);
comparison of two or more rows (places) or columns (characteristics); or
study of rows and columns together. Adding further matrices, one for
each time-period (Figure 3.3) allows for five further types of study, based
on the earlier five but concentrating on changes over time. Thus, he
concluded, systematic and regional geography are part of the same
enterprise—a repetition of Hartshorne's arguments—with neither
sufficient in itself; the revolution was not changing everything.

Berry's matrices referred only to the characteristics of places, and
further matrices could be introduced (Figure 3.4) which show the flows
between places, with one matrix for each flow category in each time-
period (Clark, Davies and Johnston, 1974). Berry used this extension
himself, though he did not formalize it, in his attempt to fuse the
procedures for formal and functional regionalization (see above, p. 35)
to produce a general theory of spatial behaviour—Berry's (1968) field
theory, which he applied in a large study of the spatial organization of
India (Berry, 1966). The techniques which he used in this work became
widely used in the 1960s, as access to high-speed computers for the
handling of very large data matrices became very easy, if not universal,
for university academics. They were given the umbrella term of factorial
ecology (Berry, 1971), and were widely applied in many aspects of

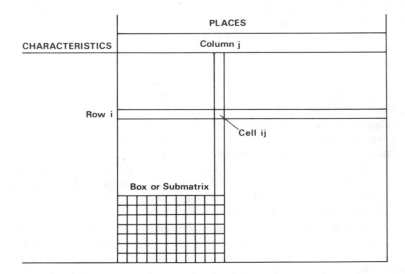

Figure 3.2   The geographic matrix: each cell—ij—contains a 'geographic
fact'—the value assumed by characteristic i at place j. Source: Berry (1964b, p.
6).

geography (e.g. Zelinsky, 1974), resulting in a methodological unity which was previously unknown across the various systematic specialisms.

The developments of the 1950s in the United States spread into several of the discipline's topical specialisms, and by the mid-1960s use of statistical methods to test hypotheses was common. It was indeed the methods which gave unity to these specialisms, which in terms of substance remained very separate, as identifiable branches within the geographical enterprise. The decline of interest in work which aimed to integrate the findings of the specialisms into regional syntheses, despite the efforts of Berry and a few others (see also Taaffe, 1974), meant that human geography experienced a centrifugal trend with regard to substance contemporaneously with a centripetal one with regard to procedures. Since the scientific method, and the statistical techniques, were used much more widely than in human geography alone, it seems that the former of these trends was probably the most important.

Transatlantic translation

By the early 1960s the quantitative and theoretical revolutions were having considerable impact beyond the United States, as the result of two agencies. The first was the publication of the work of the American iconoclasts in the major journals. American journals are widely read in Britain (much more than are British journals in America), and emulation was likely. Secondly, and probably more importantly, during the 1950s and 1960s a number of British geographers went to the United States, either as postgraduate students or as visiting staff members. Some returned with the new ideas, which they disseminated among their students and, via the Study Group in Quantitative Methods of the Institute of British Geographers, their fellow academics. (Others stayed in North America: Brian Berry was one.) There was also a local base, mainly in physical geography and arising out of the early use of statistics by climatologists (e.g. Crowe, 1936): as a result, and perhaps somewhat surprisingly, the first undergraduate text in statistics for geographers was written by an English academic (Gregory, 1963). In addition, there was some teaching of location theory—at University College, London, in the early 1950s, for example (Halvorson and Stave, 1978).

Although statistics courses were introduced in several university departments of geography in the mid-1960s, and aspects of the scientific methodology were taught in several places by at least a few staff members (Whitehand, 1970), the main focus for the introduction of the 'new geography' to Britain during the early years of the decade was the University of Cambridge. The leaders were R. J. Chorley (a geomorphologist, who had spent some time studying in the United States) and P. Haggett (a Cambridge-trained human geographer, although his early

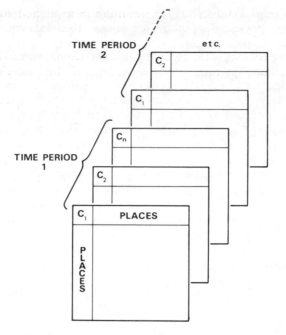

Figure 3.3   A third dimension to the geographic matrix. Source: Berry (1964b, p. 7)

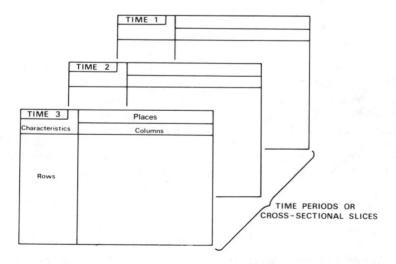

Figure 3.4   Geographic flow matrices, one per commodity, for each time period. Source: after Clark, Davies and Johnston (1974).

published work was in biogeography, who had also visited the United States and experienced the development there: Haggett, 1965c, p. vi). Their impact on British geography was considerable, through both their teaching and their publications (Gregory, S. 1976). They worked, for example, on the adaptation of certain statistical techniques to geographical (both physical and human) ends (Chorley and Haggett, 1965a; Haggett, 1964: Haggett and Chorley, 1969), but their most lasting contribution was probably the editing of two collections of papers which resulted from courses that they directed, aimed at introducing the 'new geography' to teachers.

The first of these books—*Frontiers in Geographical Teaching* (Chorley and Haggett, 1965b)—was based on a course given in 1963 which was designed 'to bring teachers and like persons into the University, there to encounter and discuss recent developments and advances in their subjects' (p. xi). In it Wrigley (1965), for example, discussed the changing philosophy of geography and saw the increasing use of statistical techniques as the contemporary development 'of singular importance' (p. 15). He pointed out that techniques of themselves do not form a methodology and that 'Geography writing and research work has in recent years lacked any general accepted, overall view of the subject even though techniques have proliferated' (p. 17). He offered no outline of a methodology, however, arguing that eclecticism in mode of analysis was likely to be most productive and that 'the best sign of health is the production of good research work rather than the manufacture of general methodologies' (p. 17). Many of the other chapters interpret geography as if the 'revolutions' had not occurred in the United States: Smith's (1965) on historical geography, for example, is an excellent British companion to the American statement published a decade earlier (Clark, 1954).

Elsewhere in the book, Pahl (1965) introduced the models of the Chicago school of urban sociologists, and suggested a social geography in which the prime factor is distance (p. 95) but it was only the chapters by Haggett and by Timms which introduced much of the transatlantic turmoil. Haggett (1965a) wrote on the use of models in economic geography, both those based on simple views of the world, such as developments of von Thunen's (Chisholm, 1962), and those derived from observations of particular cases (e.g. Taaffe, Morrill, and Gould, 1963). He noted that

> Perhaps the biggest barrier that model builders in economic geography will have to face in the immediate future is an emotional one. It is difficult to accept without some justifiable scepticism that the complexities of a mobile, infinitely variable landscape system will ever be reduced to the most sophisticated model, but still more difficult to accept that as individuals we suffer the indignity of following mathematical patterns in our behaviour (p. 109),

and as a consequence introduced the notion of indeterminacy at the individual level and showed how random variables must be introduced to operational models. His chapter on scale problems (Haggett, 1965b) illustrated methods of sampling and of map generalization from samples. Timms (1965) demonstrated the use of certain statistical techniques for the analysis of social patterns within cities (based on Shevky and Bell's social area analysis, and thus developed independently of Berry's work on this topic—see above, p. 73), pointing out that

> The sciences concerned with the study of social variation have as yet produced few models which can stand comparison with the observed patterns or which can be used to predict those patterns. . . . Prediction rests on accurate knowledge of the degree and direction of the interrelationships between phenomena. This can only be attained by the use of techniques of description and analysis which are amenable to statistical comparison and manipulation. If the goal of geographical studies be accepted as the formulation of laws of areal arrangement and of prediction based on those laws, then it is inevitable that their techniques must become considerably more objective and more quantitative than heretofore (p. 262).

If the majority of the contributors to *Frontiers in Geographical Teaching* were not as committed to the 'new geography' as was Timms (later to become a professor of sociology), this cannot be said of the editors, who used their epilogue (Haggett and Chorley, 1965) to present a strong case for the 'theoretical revolution':

> We cannot but recognize the importance of the construction of theoretical models, wherein aspects of 'geographic reality' are presented together in some organic structural relationship, the juxtaposition of which leads one to comprehend, at least, more than might appear from the information presented piecemeal and, at most, to apprehend general principles which may have much wider application than merely to the information from which they were derived. Geographical teaching has been remarkably barren of such models. . . . This reticence stems largely, one suspects, from a misconception of the nature of model thinking. . . . Models are subjective frameworks . . . like discardable cartons, very important and productive receptacles for advantageously presenting selected aspects of reality (pp. 360–1).

This view was the dominant one in their next, and substantially more influential, volume (Chorley and Haggett, 1967).

Models in Geography presented a synthesis of most of the work completed before the mid-1960s by adherents to the 'quantitative and theoretical revolutions'. Individual authors had been asked 'to discuss the role of model-building within their own special fields of research' (Haggett and Chorley, 1967, p. 19), which resulted in a series of substantive review essays, some dealing with particular topical specialisms (urban geography and settlement location; industrial location; agricultural activity—there were similar reviews for physical geography), some with particular themes ranging across several specialisms (economic development; regions; maps; organisms and ecosystems; the evolution of spatial patterns), and some with methods and approaches (demographic models; sociological models; network models). A catholic use of the term model was allowed, allowing its use as a synonym for a theory, a law, a hypothesis, or any other form of structured idea (see Moss, 1970). The approach was strongly nomothetic, however: as Harvey (1967a, p. 551) expressed it,

> the student of history and geography is faced with two alternatives. He can either bury his head, ostrich-like, in the sand grains of an idiographic human history, conducted over unique geographic space, scowl upon broad general-ization, and produce a masterly descriptive thesis on what happened when, where. Or he can become a scientist and attempt, by the normal procedures of scientific investigation, to verify, reject, or modify, the stimulating and exciting ideas which his predecessors presented him with.

All the contributors to the book had clearly chosen the latter course: their focus is on models—on generalizations of reality—and methods are very much secondary.

The orientation of this significant volume is given by the editors' introduction. (The significance lay in its two uses: first, as a synthesis and argument, the volume was widely read and used by researchers and teachers as a guide; secondly, as a series of major reviews, when repub-lished as a series of paperback volumes, the book was extensively em-ployed as an undergraduate text.) Haggett and Chorley (1967, p. 24) presented the model as:

> a bridge between the observational and theoretical levels . . . concerned with simplification, reduction, concretization, experimentation, action, extension, globalization, theory formation and explanation (p. 24).

It can be descriptive or normative, static or dynamic, experimental or theoretical (see also Chorley, 1964). It forms the basis for a paradigm, which made no attempt 'to alter the basic Hartshorne definition of Geography's prime task' (p. 38) but offered much greater progress:

> the new paradigm . . . is based on faith in the new rather than its proven ability. . . . There is good reason to think that those subjects which have modelled their forms on mathematics and physics . . . have climbed consid-erably more rapidly than those which have attempted to build internal or idiographic structures (p. 38).

Models in Geography stands as a statement of that faith, and as a major illustration of the expanding use of scientific methods in the systematic fields of human geography.

Although the editors and contributors to *Models in Geography* repre-sent most of the early active participants in the move to change British geography towards a 'more scientific' approach, there were others who are not represented, directly, in that book. Notable amongst them was a group who graduated at the University of Cambridge in the 1950s (Haggett was among them), having been tutored by A. A. L. Caesar

(Chisholm and Manners, 1973, p. xi): the group included Michael Chisholm, Peter Hall, and Gerald Manners. Only Haggett of this group undertook much work of a dominantly 'quantitative' nature: the others were more concerned with theoretical developments within systematic fields (e.g. Chisholm, 1962, 1966, 1971a).

The relatively untouched

The NAS/NRC (1965) report (see above, p. 70) indicated that the two main systematic specialisms within human geography relatively untouched by the developments were cultural and historical geography: in addition, despite Berry's attempt to reframe it (Berry, 1964b), regional geography remained largely separate from the changes in methodological emphasis. Not all cultural, historical, and regional geographers ignored the changes occurring elsewhere, of course: some, indeed, were in the vanguard of the 'revolution'—two of the chapters in *Models in Geography*, for example, were written by individuals (David Grigg and David Harvey) who showed a clear focus on historical geography in their earlier research (and later, in the case of Grigg). But in general terms the NAS/NRC report was undoubtedly correct; there is not much evidence of success in winning cultural, historical and regional geographers over to the new methodology.

Of the three groups listed in the previous paragraph, the historical geographers have probably been most concerned about their apparent isolation within the discipline. This concern has been summarized by Baker (1972) in terms of the approaches which historical geographers need to consider in greater detail:

> An assumption is necessary here: that methodologically the main advances can be expected from an increased awareness of developments in other disciplines, from a greater use of statistical methods, from the development, application and testing of theory, and from exploitation of behavioural approaches and sources. . . .
> Rethinking becomes necessary because orthodox doctrines have ceased to carry conviction. As far as historical geography is concerned, this involves a questioning of the adequacy of its traditional methods and techniques (p. 13).

All of these would have to be followed with care, and the potentials of the methodological developments assessed with caution, but Baker clearly believed there was considerable scope for change, as had already been shown in economic and in social history, and perhaps even more so in archaeology. Particular areas of historical geography, perhaps those relating to urban settlements (e.g. Ward, 1971; see also Johnston and Herbert, 1978, p. 20), are perhaps more open to such changes than are others, if for no other reason than the better quality, as well as quantity, of available data, and there is evidence of more work of this type during the 1970s (e.g. Whitehand and Patten, 1977). There are possible implications in such work, however: as Baker notes,

> Studies in, for example, 'historical agricultural geography', 'historical urban geography' and 'historical economic geography' seem to offer possibilities of fundamental development, particularly in terms of a better understanding of the processes by which geographical change through time may take place. Such an organization of the subject would view historical geography as a means towards an end rather than as an end in itself (p. 28).

Cultural geographers would appear to have been less concerned about their apparent drift away from the mainstream of geographical activity than were historical geographers, perhaps because of the lack of any parallel developments to those in geography in anthropology, the discipline with which cultural geographers probably have most contact (see Mikesell, 1967, for a general comparison). (This is a generalisation, of course. Anthropological work has experienced major paradigmatic threats, if not changes, notably in the structural work of Claude Lévi-Strauss; see Leach, 1974.) The increasing general interest of geographers in diffusion, stimulated by Hägerstrand's (1968) work and extended into other contexts, such as relatively 'underdeveloped' societies, has led to contact between the spatial analysis and cultural analysis schools of thought, however (Clarkson, 1970). Nevertheless, as Mikesell (1978, p. 1) expressed it, 'Stubborn individualism and a seeming indifference to academic fashion are well-known characteristics' of cultural geographers, whose preferences are for: a historical orientation; a focus on man's role in environmental change, on material culture, and on rural areas; links with anthropology; an individualistic perspective; and field work.

Finally, the situation with regard to regional geography has been surveyed by Paterson (1974), whose essay comprises two main sections—'On the problems of writing regional geography' and 'Is progress possible in regional geography?'—leading up to a conclusion entitled 'Regional geographers: last of the handloom weavers?'. Within the first of the major sections he investigates six problems, which include the growing shortage of subordinate materials (micro-regional studies), and the increasing submergence of regional distinctiveness, though

> only a certain amount of innovation is possible if the regional geographer is to perform his appointed task, which is to convey to his reader the essentials of his region; to illuminate the landscape with analytical light. Landforms and climate are common to all terrestrial landscapes, and human activities to most of them: how shall repetition be avoided? (p. 8). So long as contrasts between region and region (remain), and no matter to what they are attributable, there is work for the geographer to do (p. 16).

He does not conclude, despite the constraints, that there is no possibility of progress: regional geography, he argues, can advance with regard to two criteria—content and insight. Reference to Zelinsky's (1973b) *The Cultural Geography of the United States* illustrates the increased range of content currently being introduced; discussion of Meinig's work (e.g. Meinig, 1972) shows the ability of regional geographers to produce fresh spatial insights, although he has to conclude that 'Adventurousness is not a quality that most of us associate with regional geography' (p. 19). Thus:

> The way is open for regional studies which are less bound by old formulae; less obliged to tell all about the region; more experimental and, in a proper sense of the word, more imaginative than in the past, and covering a broader range of perceptions, either popular or specialist (p. 23).
>
> Regional Geography['s] . . . goals are general rather than specific; it is not primarily problem-orientated but concerned to provide balanced coverage, and its aims are popular and educational rather than practical or narrowly professional. Such relevance as it possesses it gains by its appeal . . . to the two universal human responses of wonder and concern. . . . One may recall Medawar's assertion that in science we are being progressively relieved of the burden of singular instances, the tyranny of the particular, and in turn assert that there is a frame of mind on which the particular exercises no tyranny, but a strong fascination (p. 21).

The implication to be drawn from the preceding paragraphs is that there is a difference between, on the one hand, many historical geographers and, on the other, most cultural and regional geographers with regard to the degree to which they have felt 'left behind' or 'relatively untouched' by the changes that occurred during the 1950s and 1960s in other branches of human geography. Certainly the former seem to have been impelled at least to consider the possibility of making methodological changes whereas the latter have continued to work within their established tradition (see also Mikesell, 1973). Not all historical geographers would agree with Baker's analogy from systems theory that simply 'Historical geography has a long relaxation time' (p. 11), however, and Chapter 5 indicates the degree to which they have mounted an attack against the positivist approach, as an alternative to either 'ignoring it in the hope that it will go away' or 'joining it because it can't be beaten'.

Conclusions

Reaction to the regional paradigm began to take shape in the United States during the mid-1950s. To a large extent, the aims of research in human geography were not debated, and relatively traditional definitions of the field were observed: the main issues concerned means and methods. The innovations of the period involved the strengthening of the systematic and topical geographies, and their release from a largely subservient relationship to regional geography, by attempts to develop laws and theories of spatial patterns, using models of various kinds for illumination, and by applying mathematical and, especially, statistical procedures to facilitate the search for generalizations. Whereas the regionalists saw geography as, at most, law-consuming, those of the new persuasion aimed at the production of their own laws.

These changed means to the geographic end were rapidly accepted in many branches of human geography, particularly in those topical special-isms dealing with economic aspects of contemporary life. They were soon accepted in the growing field which focuses on contemporary social geography, but were relatively ignored in historical geography and almost completely shunned in cultural and regional investigations, which, it was claimed, still focus on unique characteristics of unique places. They also spread into the corresponding fields of study across the Atlantic (and across the Pacific, too), and within little more than a decade British geographers produced a major review volume containing 816 pages of testimony to the enthusiasm of the innovators and the links which they had established with other sciences. But methods are insufficient to sustain an academic revolution unless they can be applied to a coherent substantive core. It is to the search for such a core that the next chapter turns.

4

The search for a focus

As shown in the previous chapter, the changes in human geography which emanated from several centres in the United States during the 1950s were very much concerned with methods of investigation. The ultimate aim of geographical study—as stated in Hartshorne's revised definition (p. 46)—remained the same: indeed, one could argue, from the definitions offered by Ackerman (1963) and the NAS/NRC commit- tee (1965—see above, p. 70), that human geographers had become even more ambitious in their hopes of explaining 'the world-wide system comprising man and his natural environment'. But within this general ethos, the proximate aims of geographical investigation were not always clear. Systematic studies were in the ascendant, and the intent was to develop valid laws and theories within the implicit positivist framework, but what the exact content of those laws and theories should be was not immediately apparent.

In furthering their analyses, human geographers increasingly sought a clear identity of their own within the social sciences. They required their discipline to provide a particular viewpoint and contribution to the overall goal of the group of disciplines with which they made common cause. For them, geography required a new philosophy as well as a new methodology. This new philosophy was developed around two inter- dependent foci: the spatial variable and the study of spatial systems.

Spatial variables and spatial systems

Geography is a discipline in distance, according to the inaugural lecture given by a Scottish professor (Watson, 1955), and its central theme is the relative location of people and places. The importance of relative location within society, and therefore to geography, is seen by Cox (1976) as a result of alterations in societal structure consequent upon technical change. In primitive societies, the main interactions are between rela- tively isolated groups of men and their physical environments, so that a natural early focus for geographical work was the relationships between societies and 'a spatially differentiated nature' (p. 192). With technologi- cal advancement, however, the main links are between men and other men. Interdependence within and between societies increases as a consequence of the more complex differentiation between places which reflects the division of labour, so that the most important facts in modern human existence relate to spatially differentiated societies, not to a spatially differentiated nature. It is this interdependence between groups living in different places which creates the patterns of human occupance on the surface of the earth, and provides the basic subject matter for human geographers.

The focus on spatial arrangements of spatial structures—the areal differentiation in human activities and the spatial interactions which this produces—and the role of distance as a variable influencing the nature of those arrangements can be identified in the textbooks of the 1960s which summarized for the next generation of students the contemporary activities of human geographers. A pioneer among such texts was Haggett's (1965c), whose depiction of pattern and order in spatial structures was phrased within a decomposition of nodal regions into five geometrical elements; a sixth element was added in the second edition of the book (Haggett, Cliff and Frey, 1977).

The geometrical elements in Haggett's schema (Figure 4.1) assume a spatially differentiated society within which there is a desire for interaction; people in place A want to trade with those in place B, for example, whereas those in place C want goods and services which they themselves cannot provide. This results in patterns of *movement*—of goods, of people, of money, of ideas, and so on—between places and so the first

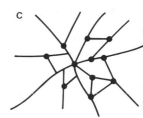
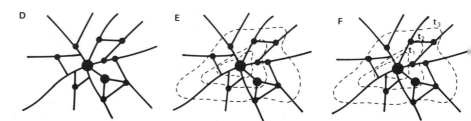

Figure 4.1   The elements in Haggett's schema for studying spatial systems: A movement; B channels; C nodes; D hierarchies; E surfaces; and F diffusion. Source: Haggett, Cliff and Frey (1977, p. 7).

element in the analysis of nodal regions involves the representation of the patterns of movement. Some movement is unimpeded—aircraft can move in all directions—but most is channelled along particular route corridors. Thus the second element in the analysis involves characterization of the movement *channels* or networks. Networks comprise edges and vertices; in a transport system, many of the latter are the *nodes*, the organizational nexus. Their spatial arrangement forms the third element in the decomposition of the nodal region, whereas the fourth investigates their organization into *hierarchies*, which define the importance of places within the framework of settlements. Finally, in the original scheme, there are the *surfaces*, the areas of land within the skeleton of nodes (settlements) and networks (routes) which are occupied by land uses of various types and intensities.

Patterns in the human occupation of the earth's surface change frequently in 'modern' societies, and the spatial order in such changes forms the sixth element in Haggett's revised schema. Change does not take place uniformly over space in most circumstances: usually it originates at one or a few localities from whence it spreads to others, along the movement channels, through the nodes, across the surfaces, and down the hierarchies. The processes of change over space and time thus involve spatial *diffusion*.

Haggett and his associates stress that geography is a science of distributions, and their emphasis is on the regularities in various elements of these distributions. Other texts have taken a similar general view of the study of what have been termed spatial systems, although perhaps emphasizing different aspects of those systems. Morrill (1970a), for example, chose a title—*The Spatial Organization of Society*—which clearly emphasizes his view of the role of geographical analysis in the larger task of the social sciences, 'understanding society'. According to him, the core elements of human geography are:

> Space, space relations, and change in space—how physical space is structured, how men relate through space, how man has organized his society in space, and how our conception and use of space change (p.3).

In this context, space has five qualities relevant to the understanding of human behaviour: (1) distance, the spatial dimension of separation; (2) accessibility; (3) agglomeration; (4) size; and (5) relative location. Put together, these can be used to build theories, such as those on which Garrison based his work (p. 54).

> Virtually all theory of spatial organization assumes that the structure of space is based on the principles of minimizing distance and maximizing the utility of points and areas within the structure, without taking the environment, or variable content of space, into account. Although the differential quality of area is interesting and its effect on location and interaction is great, most of the observable regularity of structure in space results from the principles of

efficiently using territory of uniform character. The theoretical structures for agricultural location, location of urban centres, and the internal patterns of the city are all derived from the principle of minimizing distance on a uniform plane (p. 15).

According to this view, all location decisions and decisions about the use of land are taken in order to minimize the costs of movement. The spatial approach to understanding society assumes a world in which one variable is predominant as an influence on human behaviour—distance—and it seeks to account for observed spatial patterns within this framework. And so in Morrill's book:

> the explanation of spatial structure proceeds from the deductive—what would occur under the simplest conditions—to the inductive—how local factors distort this 'pure' structure. To begin with, all the local variation may introduce is a risk of missing the underlying structure. Most theory of location therefore stresses the spatial factors—above all, distance—which interact to bring about the regular and repetitive patterns (p. 20).

Morrill's book, which by no means presented an isolated, individual view (it has been widely used as a text, and was revised and republished four years after the original edition), thus suggests an organizing framework for geographical scholarship centred on the single variable of distance. His summarizing 'theory' of spatial structures proceeds as follows:

1 Societies operate to achieve two spatial efficiency goals:
   (i)   to use every piece of land to the greatest profit and utility; and
   (ii) to achieve the highest possible interaction at the least possible cost.
2 Pursuit of these goals involves four types of location decisions:
   (i)   the substitution of land for transport costs when seeking accessibility;
   (ii) substituting production costs at sites for transport costs when seeking markets;
   (iii) substituting agglomeration benefits for transport costs; and
   (iv) substituting self-sufficiency (higher production costs) and trade (higher transport costs).
3 The spatial structures resulting from these decisions include:
   (i)   spatial land-use gradients; and
   (ii) a spatial hierarchy of regions.
   These are somewhat distorted by environmental variations to produce:
   (iii) more irregular but predictable patterns of location.
   Whereas over time distortion may result from:
   (iv) non-optimal location decisions; and
   (v)   change through processes of spatial diffusion.

Like Haggett, therefore, Morrill stresses the geometry of man's organization of his activities on the earth's surface, but whereas Haggett emphasizes pattern and geometry Morrill pays more attention to the decision-making processes which would produce the most 'efficient' pattern, as an underlying basis for the imperfect examples of that pattern which are observed in the 'real world'. Other texts (such as Abler, Adams and Gould, 1971) have followed Morrill's lead.

These textbooks illustrate the centrality of space or distance as the major focus of geographical interest in the 1960s. The emphasis on pattern, evident especially in Haggett's book, was noted by King (1969) in a major review of 'The geographer's approach to the analysis of spatial form . . . the mathematics which are used and the geometrical frameworks which are favoured' (p. 574). His focus was descriptive mathematics which would represent 'what is' rather than 'what should be'. He realized that 'when they are pursued to their extremes in very formal terms these studies run the risk of appearing as seemingly sterile exercises in pure geometry' (p. 593) but felt that geographers had not proceeded very far in providing process theories which would account for observed spatial patterns. The school of thought which he was reviewing involved working backwards, finding what order there was to explain rather than deducing what the world should like from knowledge of human behaviour, although Morrill's book was an attempt to make such deductions.

Spatial theory

Just as the methodological developments reviewed in the previous chapter proceeded without any clear guidelines in terms of a programmatic statement, so the growth of the spatial viewpoint similarly lacked any manifesto. (Watson's paper of 1955, referred to above, was not widely referenced.) The only attempt to provide such a lead—apart from general statements about geography and geometry, such as Bunge's (1962)—was provided by Nystuen (1963), in a paper which was not widely read until its reprinting in 1968.

Nystuen's objective was 'to consider how many independent concepts constitute a basis for the spatial point of view, that is, the geographical point of view' (p. 35—all page references are to the 1968 reprint) so that rather than look at the 'real world', with its many distorting tendencies, he sought clarity in considering abstract geographies. To illustrate his deduced basic concepts he used the analogy of a mosque completely lacking furniture (i.e. an isotropic plain) in which a teacher chooses a location at random. His students then distribute themselves so that they can see and hear him; their likely arrangement is in semi-circular, staggered rows facing him, with a greater density close to him. This arrangement has three characteristic features:

1   Directional orientation—they all face the teacher, the better to hear his words and to perceive his expressions;

2   Distance—they cluster around him, because the effectiveness of his voice diminishes with distance; and

3   Connectiveness—they arrange themselves in rows, so organized that each person has a direct line of sight to the teacher.

The third of these, connectiveness, is in part a function of distance and direction, but not entirely so. As Nystuen expresses it:

> A map of the United States may be stretched and twisted, but so long as each state remains connected with its neighbors, relative position does not change. Connectiveness is independent of distance and direction—all these properties are needed to establish a complete geographical point of view (p. 39).

In addition

> Connections need not be adjacent boundaries or physical links. They may be defined as functional associations. Functional associations of spatially separate elements are best revealed by the exchanges which take place between the elements. The exchanges may often be measured by the flows of people, goods, or communications (p. 39).

In the mosque, therefore, the connectivity between teacher and student involves not only a direct line of sight between them but also a flow from one to the other, in this case of ideas.

These three concepts—direction, distance, and connectiveness—are the necessary and sufficient ones for the construction of Nystuen's abstract geography, which is grounded in the study of sites (abstract places) rather than of locations (real places).

> The terms which seem to me to contain the concepts of a geographical point of view are *direction* or *orientation*, *distance*, and *connection* or *relative position*. Operational definitions of these words are the axioms of the spatial point of view. Other words, such as pattern, accessibility, neighbourhood, circulation, etc., are compounds of the basic terms. For abstract models, the existence of these elements and their properties must be specified (p. 41).

Nystuen was unsure whether these three comprised the full set of necessary and sufficient concepts for a geographical argument; boundary, he felt, might be a primitive concept and not a derivative of the basic three (see also Papageorgiou, 1969). But his general case was that arguments in human geography could be based on a small number of such concepts, a case that was implicitly accepted by much of the work in 'the spatial tradition', but which was rarely explicitly referred to.

An alternative approach to the writing of theory in human geography has been taken by Haynes (1975), based on the mathematical procedures of dimensional analysis. Five basic dimensions are defined—mass, length, time, population size, and value—and these are manipulated to indicate the validity of functional relationships, such as distance-decay equations (see p. 91), by checking their internal consistency. This approach is defended by the statement that:

> Although most quantitative geographers would probably claim to be engaged in the discovery of relationships, it appears that geography has not passed the

first stage (in the development of a science) with any degree of rigour. . . . With no clear idea of which variables are relevant and which particular characteristics in a system should be isolated, it is pragmatic to define our measurement scales with regard to a particular set of observations rather than the other way round. The method of physical science . . . is a superior system, as measurements can be interpreted exactly, different results compared, and experiments replicated (p. 66).

This is clearly a deductive approach to research. It is not as closely prescribed to a 'spatial view' as Nystuen's but has the same end in view—the derivation of a set of fundamental concepts which can form the basis for the writing of geographical theory, to be tested in the 'real world'.

These attempts at isolating human geography's primary concepts differ from most contemporary efforts at producing geographic theory which, according to Harvey (1967b, p. 212), were 'either very poorly formulated or else derivative'. Central-place theory, for example, was based entirely on postulates from economics about how people behave as 'economic men', to which the basic geographic concept of distance was added, producing a theory about the size and spacing of settlements. The attraction of central-place theory to many was indeed that geographers could apparently contribute their own basic concept to the development of theory, and need not be totally dependent on other disciplines for concepts, as the regional paradigm had perhaps suggested that they should be (see p. 43). For Harvey (1970) geography has a group of such concepts—location, nearness, distance, pattern, morphology; most of them compounds of Nystuen's basic terms—which could form indigenous elements in theory-building, allowing development of integrated social-science theories drawing concepts of equal status from all component disciplines.

An example of the development of theory involving both derivative and indigenous concepts is the field of diffusion research, which received considerable attention in the late 1960s (Brown, 1968). The basic behavioural postulate, taken from sociological research findings, was that the most effective form of communication about innovations is by word of mouth. Geographers expanded on this by introducing the effect of distance: most inter-personal communication, it was found, is between neighbours, so that information about innovations should spread outwards in an orderly fashion from the locations of the initial adopters. Pioneer work on this hypothesis had been carried out in Sweden by Hägerstrand, who introduced it to the Washington school in the 1950s, where it was taken up by Morrill (1968): Hägerstrand's own major work was made available in English translation in 1968. Much work has since been done, both on the process of diffusion and, even more, on patterns of spatial spread which, it is assumed, result from diffusion processes (see Abler, Adams and Gould, 1971, Chapter 11).

Social physics and spatial science
Of the three basic concepts identified by Nystuen, two—distance and
connectivity—received most attention from those advocating geography
as a spatial science. Direction was relatively ignored, except for some
work on migration patterns (Wolpert, 1967) which included a seminal
statement by Adams (1969). (This statement refers to direction in cardi-
nal terms. To some extent, any discussion of movement patterns which
identified destinations more precisely than in terms of distance from
origins involved directional analysis.)

By far the greatest volume of work on spatial science followed the lines
of research established by the social-physics school (p. 57). The relationship between distance and a variety of types of interaction—
migrations, information flows, movements of goods, etc.—had been
identified by several workers in the nineteenth century, such as Carey
(1858), Ravenstein (1885; Grigg, 1977), and Spencer (1892). The impact
of these early writings is not clear. McKinney (1968), for example, has
suggested that Stewart and others were unaware of the seeds of the
'gravity model' and 'population potential' ideas in the works of Carey and
Spencer, and claimed that 'current geographers could learn much'
(p. 105) from their publications. Warntz (1968) retorted that Stewart
was well aware of such writings, although McKinney in a rebuttal
pointed out that Stewart referred to them in his 1950s papers, but not
in those of the 1940s. Ravenstein's papers, on the other hand, were
clearly influential on later generations of researchers into migration
patterns.

Apart from Stewart, whose pioneering efforts were referred to in the
previous chapter, the most influential figure on social physics after
World War II was probably Zipf, who devised a 'principle of least effort'.
According to this, individuals organize their lives so as to minimize the
amount of work which they must undertake (Zipf, 1949). Movement
involves work, and so the minimization of movement is part of the
general principle of least effort. To explain this, he expanded on
Stewart's finding that with increasing distance from Princeton, fewer
students from each state attended that University. Two aspects of work
are involved in going to university: (1) the work involved in acquiring
information about the university; and (2) the work involved in actually
travelling there. Thus the greater the distance between a person's home
and Princeton, the less he is likely to know about the university and the
less prepared he will be to travel there. The validity of this distance-decay
argument was tested against many other data sets: material on the
contents of newspapers and on their circulation, for example, illustrated
the expected distance bias in flows of information, and data on movements between places showed that the greater the distance separating
them, the smaller the volume of inter-place contact.

Zipf called the regularity which he had identified the $P_1 P_2 / D$ relationships, and Stewart noted the analogy between this and the Newtonian gravity formula

$$F_{ij} = k \frac{M_i M_j}{d_{ij}^2}$$

where M_i and M_j represent the mass at places i and j respectively; d_{ij} is the distance separating i and j, k is a coefficient of proportionality (a scaling coefficient); and F_{ij} is the gravitational force between places i and j. For interaction patterns, this was rewritten as

$$I_{ij} = k \frac{P_i P_j}{d_{ij}^2}$$

where P_i and P_j are the populations of places i and j respectively, k and d are as in the previous equation, and I_{ij} is the amount of interaction between places i and j. Much work was done which involved fitting this equation to flow data, as shown in the reviews by Carrothers (1956) and Olsson (1965). In order to achieve reasonable statistical fits, the various elements of the equation had to be weighted, in the form

$$I_{ij} = f (P_i^a P_j^b d_{ij}^c)$$

where a, b, and c are weights estimated from the data set being analysed. Because different values of a, b, and c were produced in almost all studies, it was claimed that the so-called gravity model of interaction was 'an empirical regularity to which it has not yet been possible to furnish any theoretical explanation' (Olsson, 1965, p. 48: such an explanation, in statistical terms, was offered by Wilson, 1967). In effect, the influence of distance has been shown to vary from place to place, from population to population, and from context to context, in the use of the model to represent migration flows. The influence, as shown by the established connectivities, would seem to be virtually universal; what no theory has been able to account for is the variability in the strength of its impact.

It was not only in social physics, and the work of Stewart, Zipf and others, that distance was receiving attention as an important variable in the years after the Second World War. As Pooler (1977) has indicated, both economists and sociologists were becoming increasingly aware of its influence on behaviour, the former in the location theories of Weber, Hoover, Lösch and others, which stimulated the work of Garrison and his associates (p. 54), and also of Isard, and the latter in the studies of the Chicago school, out of which developed a large literature on urban residential patterns (Johnston, 1971). And so, according to Pooler:

> a number of geographers became aware of the spatial enquiries that were being undertaken in a social-science context outside their own discipline and, upon realizing their relevance to geography, proceeded to emulate them. The

appearance of the spatial tradition was prompted, not by discoveries from within geography, but by an awareness and acceptance of investigations external to the discipline. The space-centred scientific enquiries of other social sciences became paradigms for geographers, simply because those enquiries, being spatial, were seen to be of relevance by some practitioners (p. 69)

thereby fitting in with both the philosophical framework outlined by Schaefer and the 'quantitative revolution'.

In their distance-based analyses, various social scientists looked not only at the influence of distance itself but also at the meaning and measurement of that concept. Stouffer, for example, established a relationship in which migration between X and Y was accounted for not so much by the distance between them but rather by the number of intervening opportunities (Stouffer, 1940). In effect, he was measuring distance in terms of opportunities; the greater the number of opportunities available locally, the less work that has to be expended in moving to one. Others took up this flexible approach to measurement of the basic variable. Ullman (1956), for example, developed a schema for analysing commodity flows in which the amount of movement between two places was related to three factors:

1   complementarity—the degree to which there is a supply of a commodity at one place and a demand for it at the other;

2   intervening opportunity—the degree to which either the potential destination can obtain similar commodities from a nearer, and presumably cheaper, source or the potential source can sell its commodities to a nearer market; and

3   transferability—for complementarity to be capitalized it is necessary for the movement to be feasible, given channel, time and cost constraints.

This schema was not as easily fitted statistically as the 'gravity model', but its derivation from that is clear. Fitting such models also requires accepting that the influence of distance, as time and cost, varies from place to place and from time to time, as shown by Abler (1971), Forer (1974), and Janelle (1968, 1969).

All of this work on the analysis of various movement patterns was stimulated not only by its obvious relevance to the spatial science movement within geography, and its use in the development of location theories, but also by its clear applicability in forecasting contexts. The planning of land-use patterns and transport systems (especially road systems) became increasingly sophisticated, in a technical sense, during the 1950s and 1960s, first in the United States and then in Britain. Initially, data were collected to show both the traffic-generating power of various land uses and the patterns of interaction between different parts of an urban area, with the gravity model being used to describe the latter. Future land-use configurations were then designed, their traffic-generating potential derived, and the gravity model used to predict the patterns of flows and the needed road systems. Later models, most of them based on one initiated by Ira Lowry, were able to assess different configurations in terms of traffic flows, thereby suggesting the 'best' directions for future urban growth (see Batty, 1978).

The demands for sophisticated planning devices stimulated much research based on the gravity and Lowry models. American economists initiated this, but it was later taken up by British workers, led by, *inter alia*, Alan Wilson, who in 1970 was appointed to a professorship of geography at the University of Leeds, thereby becoming a professional geographer, although he had no training in the discipline. He derived the gravity model mathematically, thereby giving it a stronger theoretical base (Wilson, 1967), and developed the Lowry model into a more general set of models concerned with location, allocation and movement in space (see below, p. 109).

Much of this work involved collaboration between academic geographers and practising planners, and led to academic developments which paralleled those in regional science in the United States. Two new British journals catered for these and related research areas—*Regional Studies* and *Environment and Planning*—both of which attract contributions and readers from other social sciences. Thus the new geographical methodology was proving of considerable applied value. Careers as planners became extremely popular among geography graduates for a few years, notably in the late 1960s and early 1970s, and the growing desire of academic geographers for their discipline to equal that of other social sciences in its policy-making relevance seemed well on the way to fulfilment.

Spatial science and spatial statistics
The developments just outlined were characterized by a large volume of technical research within human geography, concerned not only with fitting the gravity model but also with a range of problems associated with the description of spatial patterns and connectivities (for example, Haggett and Chorley, 1969). Although Bunge (1962), Harvey (1969a) and others argued that the language for the analysis of spatial form is geometry, much of this work in fact concentrated on the application of descriptive and inferential statistics to spatial problems, in line with the strong empirical tradition in geographical work.

Most researchers, including Garrison (1956a), accepted that statistical procedures developed for other fields of enquiry could be adopted for geographical investigations without difficulty. For some, it was necessary to introduce modifications—as in spatial sampling (Berry and Baker, 1968)—but, despite the early work of Robinson (1956; see p. 53), it was generally assumed that there were no technical problems involved in the application of standard procedures to spatial data sets. A number of student texts appeared, especially in the late 1960s, and these made little or no reference to any peculiarities of geographical data; they differed from the texts produced by other social scientists only in the nature of their examples.

During the late 1960s, however, a group of researchers at the University of Bristol began to question the widespread assumption about the relevance of most statistical procedures in geographical investigations. For some time, statisticians had been aware of difficulties in applying the general linear model—notably in its regression and correlation form—to the time-series data used in economic analysis and forecasting. The major issue was that of autocorrelation. One of the assumptions of the model is that all observations are independent of each other; the magnitude of one reading on a variable should in no way influence that of any other. Such autocorrelation clearly exists in most time series: the value of the retail price index at one date, for example, has a very strong influence on what the value will be at nearby later dates, and is itself influenced by the value at earlier times. Because of such interdependence between adjacent observations, conventional regression methods could not be used: autocorrelation led to biased regression coefficients and therefore cast doubt on the validity of any forecasts.

The British group argued that this autocorrelation problem also existed with spatial data sets, and further that it was much less tractable than was so with temporal series. Time progresses in one direction only, but space is two-dimensional, and the independence requirement can be violated in all directions around a single point. Spatial autocorrelation, therefore, can involve all neighbours influencing all others with regard to the values of a particular variable, as recognized by some statisticians (e.g. Geary, 1954) and hinted at by geographers such as Dacey (1968) whose work stimulated the Bristol investigations. The British group expended considerable effort over a number of years on this problem (Cliff and Ord, 1973), and the research was extended over a wide range of spatial applications, including the well-known gravity model (Curry, 1972; Johnston, 1976a).

Recognition of the spatial autocorrelation problem indicated severe constraints on the application of conventional statistical procedures to geographical analyses, since the argument was that application of regression methods, and others based on them such as principal components and factor analyses, was invalid with spatial data sets. As Harvey (1969a) put it:

> The choice of the product-moment correlation coefficient for regionalization problems appears singularly inappropriate, since one of the technical requirements of this statistic is independence in the observations. Since the aim of such regionalization is to produce contiguous regions which are internally relatively homogeneous, it seems almost certain that this condition of independence in the observations will be violated (p. 347).

The general tenor of the argument was obvious for, as Gould (1970a) noted, spatial autocorrelation was the order that geographers were seeking to establish with their laws and theories. On realizing the force of the case, some accepted that conventional statistical procedures could not be applied in their work (e.g. Berry, 1973b) and the Bristol group omitted them from their rewrite of Haggett's major text (Haggett, Cliff and Frey, 1977). Most continued to apply the methods, however, either in ignorance of the autocorrelation case or on the grounds that the biased-coefficients problem refers only to the use of the general linear model in forecasting and prediction, and does not affect use of the procedures for description (see Johnston, 1978a).

The forecasting and prediction issue was basic to the work of the Bristol group, however, since their activities were focused on what they termed spatial forecasting, which involved development of procedures for estimating how trends—in the spread of a disease, for example, and in prices and unemployment—would proceed through time and over space (Haggett, 1973; Cliff *et al.*, 1975). They organized a major symposium on this (Chisholm, Frey and Haggett, 1971) and have stimulated a considerable volume of work, much of it technical, on the problems of identifying and forecasting spatio-temporal trends (e.g. Bennett, 1978b). Hägerstrand's work on diffusion patterns is the natural base for much of this research, which focuses on the patterns of spread rather than on their generating processes.

The nature of such spatial forecasting has been reviewed by Hay (1978), who challenges what he sees as its fundamental assumption, that phenomena behave coherently in both time and space. He questions the assumed stability of inter-place interrelationships over time present in analyses such as that of Martin and Oeppen (1975) on market-price variations, and argues instead for consideration of the value of catastrophe theory. In this, small changes in the control variable(s) can lead to major changes in the dependent variable being studied (such major changes are the catastrophes). If catastrophes occur, then the linear extrapolations typical of the work of the Bristol group have clear limitations as forecasting procedures; as yet, relatively little work has been done by geographers on the application of this relatively new area of mathematics to their problems (Wilson, 1976a).

One final statistical, and substantive, problem in the spatial approach has been identified by Harvey (1973, p. 40); he calls it the problem of confounding. This involves a non-experimental situation in which there are two interdependent variables operating on a third, and it is extremely difficult to establish the relative importance of either. Thus, for example, geographical studies show that city residents tend to marry people who live nearby, whereas sociological investigations show that most marriages are between people of the same social background. Amalgamation of these two findings would suggest that people tend to choose spouses from among those of similar social backgrounds and who live in the same area as themselves. But, as Harvey (1973) expresses it:

> Now it so happens that people of the same class tend to live in proximity to one another. How then are we to distinguish how much the spatial variable contributes, and how much the personality variables contribute? . . . We do not appear to possess adequate non-experimental research designs to allow us to handle this sort of problem in any but the crudest fashion (pp. 40–1).

Does the spatial influence produce the social-class effect? Does the social-class effect—and the associated spatial segregation of classes—produce the distance effect? Or are both operating, and in what relative strength? The problems of answering these questions raise important doubts, at least in some minds, about the significance of the spatial point of view to the study of human behaviour.

Countering the spatial separatist theme

Opposition to the widespread view of human geography as a spatial science developed largely as a counter to its claims for separate status for such a discipline. Some, such as Crowe (1970), saw the use of the spatial variable in nomothetic studies as a naive spatial determinism which paralleled the earlier environmental determinism. Others based their criticisms on the implicit divisibility of the social sciences in the spatial claims, and it is these criticisms which form the subject of the present section.

The most sustained argument against the view that geography is a spatial science—what he calls the 'spatial separatist' theme—has been presented in a series of papers by Sack, a former associate of Lukermann at the University of Minnesota (see p. 68). There are three dimensions of reality—space, time, and matter—and geography, according to the spatial-separatist view, is the science of the first of these. But according to Sack, space, time and matter cannot be separated analytically in an empirical science which is concerned to provide explanations. Thus in his first paper he set out to show that geometry is not an acceptable language for such a science (Sack, 1972). Geometry is a branch of pure mathematics which is not concerned with empirical facts; its laws are static laws, with no reference to time, and they are not derivable from any dynamic or process laws (p. 67). Geographic facts have geometric properties (locations) but if, as Schaefer proposed, geographical laws are concerned only with the geometries of facts, then they will provide only incomplete explanations of these facts. (To illustrate this contention, Sack used an analogy of chopping wood. If the answer to the question 'why are you chopping wood?' is 'because the force of the axe on impact splits the wood' then it is a static, geometric law, but if the answer is 'to provide fuel to produce heat' then it is an instance of a process law, which incorporates the geometric law. In this analogy, a process law is equated with the intention behind an action.) The laws of geometry, according to Sack, are sufficient to explain and to predict geometries, so that if geography aimed only to analyse points and lines on maps it could be an independent science using geometry as its language. But 'We do not accept the description of the changes of its shape as an explanation of the growth of a city' (p. 72) so that 'Geometry alone, then, cannot answer geographic questions' (p. 72) leading to the conclusion that:

> To explain requires laws and laws (if they are valid) explain events. Since the definition of an event implies the delimitation of some geometric properties (all events occur in space), the explanation of any event is in principle an explanation of some geometric properties of events (p. 77).

Thus geography is closely allied with geometry in its emphasis on the spatial aspects of events (the instances of laws), but geometry alone is insufficient as a basis for explanation and prediction since no processes are involved in the derivation of geometries.

Bunge (1973a) responded to this statement, claiming that spatial prediction was quite possible with reference to the geometry alone, as instanced by central-place theory and Thunian analysis. Such geometries provide 'classic beauty', and 'purging geometry from geography reduces our trade to no apparent gain' (p. 568). Sack's (1973a) reply was that the static laws espoused by Bunge are only special cases of dynamic laws having antecedent and consequent conditions, and that:

> Although the laws of geometry are unequivocally static, purely spatial, non-deducible from dynamic laws, and explain and predict physical geometric properties of events, they do not answer the questions about the geometric properties of events that geographers raise and they do not make statements about process (p. 569).

Sack did not argue, as Bunge supposed, that geometry should be purged from geography, but only that space should not be considered independently from time and matter. He developed this theme (Sack, 1973b) with the contention that:

> for a concept of physical space to conform to the rules of concept formation and be useful in a science of geography every instance of the geometric or spatial terms must be connected or related to one or more instances of non-geometric terms (to be called substance terms) (p. 17).

Thus physical distance is not a concept in itself: it is necessary to know the terrain which a road crosses, for example, in order to assess the significance of its length in a gravity model—geometry alone is not enough (hence the considerations of the meaning of distance referred to above—p. 92—and the work of Mackay (1958) on the influence of boundaries on movement patterns). Since there is no such thing as empty physical space so there are no frictions of distance per se. There are frictions which demand work in crossing a substance, but it is the substance itself which creates the frictions, and the context in which it is being crossed, not simply the distance: 'There are frictions and there are distances, but there is no friction of distance' (p. 22). Geography, according to Sack, is concerned to explain events and so it requires substantive laws: such laws may contain geometric terms, such as the frictions of crossing a certain substance, but these terms of themselves are insufficient to provide explanations.

The spatial separatist approach proposes an independent position for geography within the social sciences based on its use of geometry, but Sack (1974b) contends that 'The spatial position's aim of prying apart a subject matter from the systematic sciences by arguing for spatial questions and spatial laws does not seem viable' (p. 446). Instead, two types of law relevant to geographical work must be identified (Sack, 1974a). *Congruent substance laws* are independent of location: statements of 'if A then B' are universals which require no spatial referent. *Overlapping substance laws*, on the other hand, involve spatial terms, as in central-place theory: 'if A then B' in such cases contains some specific reference to location. Both types are relevant and necessary in providing the answers to geographical questions, so no case can be made for a necessary 'spatialness' to the substance laws of human geography.

May (1970) has also argued against Schaefer's claim that geography is the study of spatial relations:

> If we extend Schaefer's argument to include time, and assign the study of temporal sequences or relations to the historian, then the only conclusion respecting this matter that can be drawn is that economic, social, political, and other relations must be non-spatial and non-temporal. Hence economics, sociology, political science, etc. are non-spatial, non-temporal sciences. But this is absurd . . . insofar as economics qualifies as a science possessing empirical warranty, then its generalizations must apply to given spatio-temporal situations (p. 188).

If, then, all sciences have a spatial content, what is there left with which to define a separate discipline of geography? May lists five possibilities.
1 Geography is a 'super-science' of spatial relations, 'a generalizing science of spatial relations, interactions, and distributions' (p. 194) drawing on the findings of other sciences. This would leave the latter truncated and their studies unfinished; in any case such an approach has been unsuccessful, 'the issue of the conception of geography as a generalizing or law-finding science that somehow stands above the social sciences and history is not even appropriately debatable' (p. 195).
2 Geography is a lower-level science of spatial relations, applying in empirical contexts the laws of higher-level, generally more abstract sciences. (This may be a description of much of the geography of the 1960s.) This again seems to truncate the latter sciences and raises the question 'what differentiates economic geography from economics, and vice versa?'.

3 Geography is the study of geographical spatial relations. This implies that there are spatial phenomena not studied in the other social-science disciplines and which can therefore be claimed as geography's: May can conceive of no objects which are purely geographical (or, in the parallel argument, purely historical either).

4 Geography is the study of 'things in reality' spatially. Yet again, this abstracts from other sciences, although May does admit that there are certain 'bits and pieces' which are not studied elsewhere; these, however, do not make a satisfactory empirical foundation for a separate discipline.

5 'Geography is not a generalizing or law-finding science of spatial relations' (p. 203).

The first four of these possibilities indicate that, because of the analytical indissolubility of time, space and matter, all social sciences are concerned with spatial relations. For May, as for Sack, therefore, geography cannot claim an independent status on the basis of the spatial variable. Moss (1970) reached a similar conclusion at about the same time:

> geometrical relationships must be assigned economic, social, physical, or biological meaning before they can in any sense become explanatory . . . though geometries may be important tools in geographical study and research, they cannot be a source of theory since their analogy with geographical phenomena is simply through particular logical structures, and not through explanatory deduction . . . such an application implies that space, area, distance, etc., are important in and of themselves, quite independently of any implications they may have in terms of diffusion, of cost, of time, or of process. This is manifestly false (p. 27).

and Gregory (1978b) has criticized the extremely narrow, even superficial view of spatial processes which he perceives in the work of Haggett and others (p. 95). Their view, he claims, is an instrumentalist one involving theories which cannot be validated conclusively but which can only be evaluated pragmatically against the real world. Bennett (1974), for example, accepts that his models do not mirror actual processes, but makes assumptions that they do allow policy formulation (and therefore produce self-fulfilling prophecies).

These arguments reviewed in the present section are highly critical of much of the work undertaken by geographers in the fashion which dominated the 1960s, impelled by the 'Victorian myth of the supremacy of the natural sciences' (Gregory, 1978a, p. 21). The alternatives to such work are discussed in the next chapter.

Systems

The study of systems, as currently understood, was first introduced to the geographical literature by Chorley in 1962 although, as Foote and Greer-Wootten (1968) noted, systems analysis was promoted in Sauer's

(1925) programmatic statement *The Morphology of Landscape* with the words 'objects which exist together in the landscape exist in interrelation': Garrison's (1960a) review paper made a similar point. To some extent, the adoption of a systems approach involved putting 'old wine into new bottles', although the holistic approach currently espoused differs very much in format from that used by Sauer. More generally, the notion of a system has a long history, as Bennett and Chorley (1978, pp. 11–14) point out: teleological traditions, for example, postulate the world as 'a vast system of signs through which God teaches man how to behave' (p. 12), whereas functionalism links observed phenomena together as 'instances of repeatable and predictable regularities' of form.

The keystone of the study of systems is connectivity. As Harvey (1969a, p. 448) points out, reality is infinitely complex in its links between variables, but systems analysis provides a convenient abstraction of that complexity in a form which maintains the major connections. In brief, a system comprises three components (p. 451):

1   A set of elements;
2   A set of links (relationships) between those elements; and
3   A set of links between the system and its environment.

The last components may be non-existent, in which case the system is termed a closed one. Closed systems are extremely rare in reality, but are frequently created, either experimentally or, more usually in human geography, by imposing artificial boundaries, in order to isolate the salient features of a system. Thus, just as an internal-combustion engine comprises a set of linked elements which receives energy from its environment and returns spent fuel to that environment, so a set of settlements linked by communications networks forms a spatial system, with links to settlements outside the defined area of the system being the contacts with the environment.

The use of systems terminology to describe a spatial assemblage was widely adopted by human geographers in the 1960s, and it formed the basis, for example, of Haggett's (1965c) pioneering text. Not all applications of the terminology involved the study of explicitly spatial systems. In many cases, the elements were phenomena whose spatial locations were not studied, and the links between the phenomena were functional relationships. In such contexts, geographical study of causal systems was identical to the study of similarly structured problems in other disciplines, but that the phenomena occupied locations and that the links involved crossing space suggested the extra, geographical dimension which led to investigations of causal, spatial systems.

The early literature on systems analysis in geography was programmatic rather than applied; it suggested how the terminology might be applied in a research and teaching context, often reinterpreting old material (McDaniel and Hurst, 1968). Relatively few applications were reported, and more than a decade later, much of the literature assessed in a major review was written by other scientists (Bennett and Chorley, 1978). Nevertheless, Harvey (1969a) wrote that:

> If we abandon the concept of the system we abandon one of the most powerful devices yet invented for deriving satisfactory answers to questions that we pose regarding the complex world that surrounds us. The question is not, therefore, whether or not we should use systems analysis or systems concepts in geography, but rather one of examining how we can use such concepts and such modes of analysis to our maximum advantage (p. 479).

In the search for answers to this question, two variants on the systems theme have been employed. The first is *systems analysis*; the second is *general systems theory*, which is an attempt to provide a more unified science than current disciplinary boundaries allow. The two are sometimes confused, but will be treated separately here.

Systems analysis

Having defined a system of geographical interest, with all the attendant difficulties that this entails (Harvey, 1969a, pp. 455–9), how might it be studied? Several typologies of systems and systems analyses have been suggested.

Chorley and Kennedy (1971) identify four types of system (Figure 4.2). *Morphological systems* are statements of static relationships—of links between elements: they may be maps showing places joined by roads, or they may be equations describing the functional relationships between variables. Much of the spatial analysis described earlier in this chapter outlined such morphological systems. *Cascading systems* contain

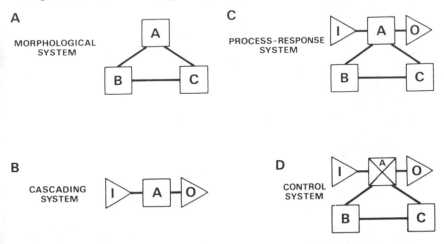

Figure 4.2 Types of system. In the diagrams, A, B and C indicate elements in the system, I = input, O = output, and in the control system, A is a valve.
Source: Chorley and Kennedy (1971, p. 4).

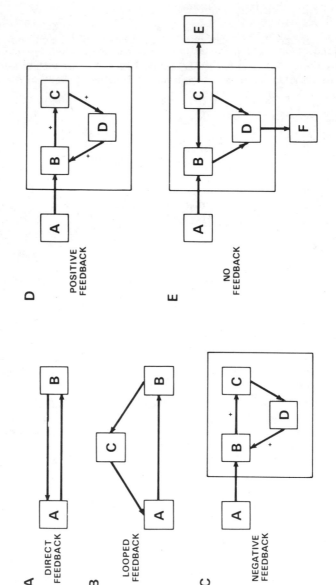

Figure 4.3   Various types of feedback relationship within systems.   Source: Chorley and Kennedy (1971, p. 14).

links through which energy passes from one element to another: factories can be portrayed as cascading systems, in that the output of one factory is in many cases the input for another. Each element may itself be a system (linked departments within a factory, for example), so a nesting hierarchy of cascading systems can be described, as with Haggett's (1965c) nodal regions and the input-output matrix representation of an economy (Isard, 1960): Berry (1966) has linked these two examples of cascading systems together in his inter-regional input-output study of the Indian economy. Within each element in a cascading system, the material flowing through is manipulated in some way (the industrial process in a factory, for example). The nature of the manipulative process may be ignored entirely in the investigation, with focus on the inputs and outputs only: in such cases the representation of the element is termed a black box. White-box studies investigate the transformation process, whereas grey-box analyses make a partial attempt at their description.

Process-response systems are characterized by studies of the effects of linked elements on each other. Instead of focusing on form, as in the first two types, these are studies of processes, of causal interrelationships. In systems terms these may involve, for example, the effects of a variable X on another, Y; in the analysis of spatial systems they could involve the effect of variable X in place a on variable Y in place b, as with the effect of inflation in the United States on unemployment in the United Kingdom at some later date, perhaps, or with the transmission of a disease from one area of a country to another (Cliff *et al.*, 1975). Finally, there are *control systems*, which are special cases of process-response systems, having the additional characteristic of one or more key elements (values) which regulate the system's operation and may be used to control it.

Attention has focused on the last two types. Langton (1972), for example, has suggested that process-response systems provide an excellent framework for the study of change in human geography. He identifies two sub-types. *Simple-action systems* are unidirectional in their nature: a stimulus in X produces a response in Y, which in turn may act as a stimulus to a further variable, Z. Such a causal chain is merely a reformulation of 'the characteristic cause-and-effect relation with which traditional science has dealt' (Harvey, 1969a, p. 455); in another language (see p. 67) it is a process law.

More important, and relatively novel to human geography, is the second sub-type, *feedback systems*. According to Chorley and Kennedy (1971):

> *Feedback* is the property of a system or sub-system such that, when change is introduced via one of the system variables, its transmission through the structure leads the effect of the change back to the initial variable, to give a circularity of action (pp. 13–14).

Feedback may be either direct—A influences B which in turn influences A (Figure 4.3A)—or it may be indirect, with the impulse from A returning to it via a chain of other variables (Figure 4.3B). With *negative feedback* the system is maintained in a steady state by a process of self-regulation known as homeostatic or morphostatic: 'A classic example is provided by the process of competition in space which leads to a progressive reduction in excess profits until the spatial system is in equilibrium' (Harvey, 1969a, p. 460). But with *positive feedback* the system is characterized as morphogenetic, changing its characteristics as the effect of B on C leads to further changes in B, via D (Figure 4.3D).

The concept of feedback, with the associated notions of homeostasis and morphogenesis, gave 'the nuclei of the systems theory of change' (Langton, 1972, p. 145): as a consequence, Langton argues that the nature of feedback should be the focus of geographical study. In many spatial systems, feedback may be uncontrolled, but others may include a regulator, such as a planning policy (Bennett and Chorley, 1978). There are few geographical studies of such feedback processes, however. For homeostatic systems, Langton cites investigations of central-place dynamics in which the pattern of service centres is adjusted as the population distribution changes, to reproduce the previous balance between supply and demand factors (e.g. Badcock, 1970). Morphogenetic systems are illustrated by Pred's (1965b) model of the process of urban growth, in which expansion in sector A generates, via a series of links, further expansion in that sector, as in Myrdal's (1957) more general theory of cumulative causation: such systems modelling has been used to predict urban futures (Forrester, 1969). But in most of these studies the input from systems theory is slight, leading Langton (1972) to the paradoxical conclusions that:

> *First*, there is little correlation between the extent of the penetration of the terminology of systems theory and the rigorous application of its concepts. The 'empty' use of terminology, which is typified by the use of the term feedback as an explanatory device rather than as a description of a fundamental research problem, must be counter-productive in a situation in which the terms themselves may be given many subtle different shades of meaning. . . .
> *Second*, somewhat paradoxically, many of the concepts of systems theory are already used in geography without the attendant jargon and without apparently drawing direct inspiration from the literature of systems theory (pp. 159–60).

This second conclusion suggests that a discipline should develop the necessary conceptual framework itself without the introduction of relevant concepts from related disciplines, but Langton disputes this, arguing that systems theory clarifies questions in established theories, focuses directly on the processes of change, and forces careful analytical study.

One of the most substantial attempts to apply systems theory to a problem in human geography is the work of Bennett (1975) on the dynamics of location and growth in northwest England. Having represented the system—its elements, links, and feedback relationships—

he estimated the influence of various external (i.e. national) events on the system's parameters, isolated the effects of government policy (Industrial Development Certificates) on the system's structure, and produced forecasts of the region's future spatio-temporal morphology. He has developed the forecasting aspects of this methodology in later papers (Bennett, 1978a, 1979) and demonstrated the relevance of the systems approach for the evaluation of policy alternatives and the understanding of very complex patterns of human decision-making.

One type of study which has firmly adopted the systems approach covers the border area between human and physical geography. The *ecosystem* is a process-response system concerned with the flows of energy through biological environments, most of which include, or are affected by, man. It is also a control system in that the living components act as regulators of the energy flows: 'they further represent a major point at which human control systems must intersect with the natural world' (Chorley and Kennedy, 1971, p. 330). Most naturally occurring ecosystems are, for much of the time, homeostatic (see Chapman, 1977, Chapter 7), but human entry often transforms them into morphogenetic systems, with potentially catastrophic effects.

Stoddart (1965, 1967b) has argued that the ecosystem should be employed as a basic geographical principle, but despite other programmatic statements (e.g. Clarkson, 1970), including two based on the associated concept of community (Morgan and Moss, 1965; Moss and Morgan, 1967), Langton's conclusion about relatively little substantive research would appear to be valid (see Grossman, 1977). Similar attempts, but involving less consideration of the biotic environment, have been made in allied disciplines, and have occasionally been imported into the geographical literature. The human ecosystem models of sociologists (e.g. Duncan, 1959; Duncan and Schnore, 1959) have been used as frameworks for the investigation of migration (Urlich, 1972) and of urbanization (Urlich Cloher, 1975), for example, and the operational research techniques of economists, with their important feedback mechanisms, have stimulated some work in transport geography (e.g. Sinclair and Kissling, 1971), but the overall impact has been slight.

The most comprehensive attempt to forge a systems approach to geographical study has been provided by Bennett and Chorley (1978). The intention of their book is to provide 'a unified multi-disciplinary approach to the interfacing of "man" with "nature" ' (p. 21), with three major aims:

> First, it is desired to explore the capacity of the systems approach to provide an inter-disciplinary focus on environmental structures and techniques. Secondly, we wish to examine the manner in which a systems approach aids in developing the interfacing of social and economic theory, on the one hand, with physical and biological theory, on the other. A third aim is to explore the

> implications of this interfacing in relation to the response of man to his current environmental dilemmas. . . . It is hoped to show that the systems approach provides a powerful vehicle for the statement of environmental situations of ever-growing temporal and spatial magnitude, and for reducing the areas of uncertainty in our increasingly complex decision-making arenas (p. 21).

This mammoth task (undertaken in a mammoth book of 624 pages!) involves elucidation of not only the 'hard systems' of physical and biological sciences but also the 'soft systems' characteristic of the social sciences. With regard to the latter, they cover a very large and fertile literature concerned, first, with the cognitive systems describing man as a thinking being and the decision-making systems used by humans, as individuals and in groups (this material is covered in the following chapter) and, secondly, with the socio-economic systems made up of very many of these interacting individuals and groups. They then attempt to interface the two types, because:

> in large-scale man-environment systems the symbiosis of man as part of the environment of the system he wishes to control introduces all the indeterminacies of socio-economic control objectives. . . . In particular, we need to ask what are the political and social implications of control, and for whom and by whom is control intended? (p. 539)

Not surprisingly, they end their essay into the field with a discussion of the very many substantial problems involved in such interfacing, although many of these are firmly tackled in the text (which was written after the widespread introduction to the literature of many of the criticisms discussed in the next two chapters).

Systems theory, information, and entropy
The definition of a system presented so far in this chapter is of a series of linked elements interacting to form an operational whole. This has been challenged by Chapman (1977) who opens his book with the statement:

> I do not think that the concept of a system will have any great operational consequences in geography for a long time yet. It represents an ideal that the real world does not fully approach. On the other hand, in conceptual terms I think the concept is extremely important and useful, and that it has a great and immediate role to play for those who are about to plan the strategy of their research. As a framework for analysis, it has no current peers (p. 6).

For Chapman, a system comprises a series of elements which can take alternative states, and his definition—following Rothstein (1958)—is

> A system is a set of objects where each object is associated with a set of feasible alternative states: and where the actual state of any object selected from this set is dependent in part or completely upon its membership of the system. An object that has no alternative states is not a functioning part but a static cog (p. 80).

An example of such a system is a number of farms, each comprising a series of fields: every farmer has to decide how to use each of his fields. In part, each of his decisions will reflect his general operations and the uses to which he puts all of his other fields; in part, too, it will be a function of the external market and the decisions made by other farmers regarding their own fields. Thus there are a large number of alternative states for the system of fields—different configurations of land uses. According to Chapman, systems analysis should involve investigation of these configurations and the placing of the observed pattern within the context of the alternatives:

> to theorize merely about what does exist is not very useful. If we restrict ourselves to that alone, all explanation will be merely historical accidental. At all stages it is most important to include consideration of what else could have been. The definition of organization in a system even explicitly requires the assessment of what else could have been (pp. 120–1).

Taylor and Gudgin (1976) have argued in the same way in a particular research context: instead of simply asking 'is there a bias in the electoral districting of a borough?' they ask 'what is the likelihood of a bias occurring, given the constraints of the system?'.

Such analyses focus on one configuration of the system as a sample from a set of possible configurations. One of their key concepts is that of *entropy*. In general terms, the entropy of a system is an index of uncertainty. (In the second law of thermodynamics an increase in system entropy is an increase in system uncertainty. A good example of this is the introduction of a layer of hot water onto a body of cold water. Initially, the two types are separate, and one can be completely certain about the location of the hot molecules, for example. But with no external influence the two slowly mix, until all are at the same temperature. As the mixing proceeds, so the entropy increases.)

Social scientists have drawn their usage of entropy from two separate, though linked, definitions. Thermodynamic entropy relates to the most probable configuration of the elements, within the constraints of the system's operations. In information theory, entropy refers to the distribution of the elements across a set of possible states, and is an index of element dispersion. One can be completely certain about a distribution, in terms of predicting where one element will be, if all elements are in the same state; conversely, one will be most uncertain when elements are equally distributed through all possible states, so that prediction of the location on any one element is most difficult. (An example of this use is in Johnston, 1976a; see also Webber, 1977, on the relationship between entropy and uncertainty.)

Used in its simplest form, the information-theory measure of entropy is another descriptive index, but it can be developed in a variety of ways. Chapman (1977) illustrates three uses:

1 as a series of indices of variations in population distributions;
2 as an index of redundancy in a landscape, where redundancy is defined as relating to a regular sequence so that it is possible to predict the land use at place a, for example, from knowledge of the land uses at neighbouring places; and
3 as a series of measures of reactions to situations in states of uncertainty.

In general terms, although not in detailed methodology, these continue the tradition of Stewart and Warntz's macrogeography (p. 58): the aim is to describe a pattern rather than to explain it, although the nature of the constraints used to derive the entropy measures provides an input to explanatory modes of analysis (Webber, 1977).

The usage of entropy as developed in statistical mechanics rather than in information theory was introduced to the geographical literature by Wilson (1970): in many ways the aim is the same as in the information-theory usage. Wilson's initial example was a flow matrix. The number of trips originating in a series of residential areas is known, as is the number ending in each of a series of workplace areas, but the entries in the cells of the matrix—which people move from which residential area to which workplace area—are unknown. What is the most likely flow pattern? Even with only a few areas and relatively small numbers of commuters, the number of alternatives is very large. Wilson defines two states of the system. The first is the macro-state, comprising a particular flow pattern: five people may go from zone A to zone X, for example, and three from the same origin to zone Y, but it is not known which five are in the first category and which three in the second. A micro-state, on the other hand, is a particular example of a flow pattern—one of the many possible configurations of eight people moving from zone A, five to X and three to Y. Entropy-maximizing procedures find that macro-state with the largest number of micro-states associated with it; in other words it is the macro-state about which the analyst is least certain of which configuration produces the pattern. Such a procedure shows that

> the most probable distribution is that with the greatest number of micro-states giving rise to it. Thus the distribution corresponds to the position where we are most uncertain about the micro-state of the system, as there are the largest possible number of such states and we have no grounds for choosing between them (p. 6).

This too follows the macrogeography tradition. It is not an attempt at explanation. Rather Wilson sees his work as illustrating

> the application of the concept of entropy in urban and regional modelling, that is, in *hypothesis development*, or theory building. ('Model' and 'hypothesis' are used synonymously, and a theory is a well tested hypothesis.) . . . the entropy-maximizing procedure enables us to handle extremely complex situations in a consistent way (pp. 10–11).

The hypothesis that the entries in the flow matrix conform to the most likely distribution can be tested against 'real' data. If it is falsified, either entirely or in part, it can be refined by building in more constraints. Wilson does this with his intra-urban transport models, for example, by introducing travel-cost constraints, different types of commuters (class, age etc.), different types of jobs, and so on. The aim is to describe the most likely system structure from a given amount of information which is incomplete.

Wilson has developed both the theory of his modelling and the substantive applications. His general text (Wilson, 1974) contains a whole family of models that can be used to represent, and then to forecast, the various components of a complex spatial system such as an urban region. Some of these models have been expanded (e.g. Rees and Wilson, 1977, on demographic accounts) and they have all been applied, with varying success, to the West Yorkshire region (Wilson, Rees and Leigh, 1977).

Wilson's work has been applied and developed by others (see Batty, 1976, 1978, for example). In reviewing his 1970 presentation, Gould (1972) termed it 'the most difficult I have ever read in geography' but continued 'he has planted a number of those rare and deep concepts whose understanding provides a fresh and sharply different view of the world' (p. 689). Webber (1977) has extended Wilson's argument that the purpose of entropy-maximizing models is to draw conclusions from a data set which are 'natural' in that they are functions of that data set alone and contain no interpreter bias. For him, they provide a convenient way of organizing processes of thinking about a complex world, and he identifies an 'entropy-maximizing paradigm' (p. 262) which focuses on location models (the probability of an individual being in a particular place at a particular time), on interaction models (the probability of a particular trip occurring at a particular time), and on joint location/interaction models.

In emphasizing aggregate patterns, this work is macrogeographic: according to Webber (1977) 'The entropy-maximizing paradigm asserts . . . that, though the study of individual behavior may be of interest, it is not necessary for the study of aggregate social relations' (p. 265). The patterns predicted by the models are functions of the constraints, so that knowledge of these means that 'the entropy-maximizing paradigm is capable of yielding meaningful answers to short-run operational problems' (p. 266) and thus is of immense value for immediate planning purposes. But:

> in the longer run, much of the economic system is variable: the constraints and the spatial form of the urban region may change . . . the research task facing entropists is (1) to identify the constraints which operate upon urban systems, which is partly an economic problem; (2) to deduce some facets of the economic relations among the individuals within the system from the use of the formalism; and (3) to construct a theory which explains the origins of

the constraints. Only when the third task has been attempted may the paradigm be adequately judged (p. 266).

Thus the entropy-maximizing model acts not only as a 'black box' forecasting device (p. 102) but also as a hypothesis: if the operation of the systems described is to be understood, the axioms—the constraints—must themselves be explained. Given the nature of the constraints (in Wilson's initial example, why people live where they do, why people work where they do, and why they spend a certain amount of time, money and energy on transport) the task is a major one: entropy-maximizing models aim to clarify it and indicate the most fruitful avenues for investigation.

The theory of systems and general systems theory

Like regional science, the development of General Systems Theory (GST) has been very much tied up with the academic career of one man—in this case Ludwig von Bertalanffy (see von Bertalanffy, 1950). It reflects an attempt to unify science via perspectivism instead of the more usual division of science through reductionism. Its focus is the isomorphisms, the common features between the systems studied in different disciplines, and 'Its subject matter is the formulation and derivation of those principles which are common for systems in general' (Walmsley, 1972, p. 23). The goal is a metatheory with rules that apply in a variety of contexts; application (which might be termed the theory of systems rather than GST) is usually by analogy from one discipline in order to advance understanding in another (as in Chappell and Webber's, 1970, use of an electrical analogue of spatial diffusion processes). For geography, GST offers an organizing framework; GST itself is an empirical exercise using inductive procedures to fashion general theories out of the findings of particular disciplines.

It has been claimed that there have been no advances in either the theoretical base or the empirical application of GST (Greer-Wootten, 1972). It has been employed by some geographers, however: Woldenberg and Berry (1967) drew analogies between the hierarchical organization of rivers and of central-place systems, for example; Berry (1964a) argued that cities are open systems in a steady-state, as exemplified by the stability of their behaviour-describing equations; and several authors (e.g. Ray, Villeneuve and Roberge, 1974) have applied the concept of allometry—that the growth rate of a component of an organism is proportional to the growth of the whole—in several contexts.

According to its proponents, the advantages of GST to human geography lie in its interdisciplinary approach, its high level of generalization, and its concept of the steady-state of an open system (Greer-Wootten, 1972; Walmsley, 1972), but they also contend that geography's strong empirical tradition means that it has more to contribute than to take from

GST. But one critic has claimed that 'General systems theory seems to be an irrelevant distraction' (Chisholm, 1967, p. 51), an argument based largely on a paper by Chorley (1962) referring to the Davisian system of landscape development. Chisholm summarizes the case for GST as:

1 there is a need to study systems rather than isolated phenomena;
2 there is a need to identify the basic principles governing systems;
3 there is value in arguing from analogies with other subject matter; and
4 there is a need for general principles to cover various systems.

In his view, however, something as grand as a metatheory is unnecessary in order to convince people of the need to understand what they study, of the value of interdisciplinary contact, and of the potential fertility of arguing by analogy.

Conclusions

As stressed in the previous chapter, much of the force of the developments in human geography during the 1950s and 1960s related to methodology and improvement of the approach to traditional geographical questions. In addition, however, attempts were made to inaugurate and press a particular geographical point of view; the two main themes advanced have been reviewed here. The first—the spatial-science viewpoint—was fairly widely and rapidly accepted and many geographers placed the spatial variable at the centre of their research efforts: as will be indicated in the next chapter, their work came under increasing attack after the mid 1960s. The second—the systems approach—has received much less detailed attention, despite frequent gestures of approval towards it. Compared with the spatial-science view, which could be rapidly assimilated within the developing statistical methodology (although note the critique discussed on p. 94), the systems approach was technically much more demanding, and perhaps for that reason attracted fewer active researchers. These have remained the source of fertile ideas, however, and have continued to publish major works throughout the 1970s, paralleling, and in some cases (e.g. Bennett and Chorley, 1978) responding to, the academic movements outlined in the following chapters.

5

Behavioural geography and alternatives to positivism

The changes outlined in the previous two chapters had been widely adopted by the mid 1960s, particularly in the United States, in Canada, and in Great Britain, but also in those other countries whose academic traditions and life are closely tied to the North Atlantic realm. As already documented, the changes were, not surprisingly, resisted by defenders of previous methodologies and philosophies. Almost contemporaneous with such defences were attacks on the positivist approaches, attacks based not on the growing split between the 'old' and the 'new' in human geography but rather on the perceived failures of the latter to provide viable modes of understanding. Some of the attacks came from without, from those who had neither accepted nor operated the tenets of positivist spatial science, but in part also they came from within, from those who had tried the 'new' and were disappointed with their 'accomplishments'. There was some contact between the two groups, including cross-referencing of literature, but their arguments led in different directions. The initial attack from within led to a modification of approach to theory, but not method, whereas the attack from without suggested new methodologies and philosophies. Separate reviews of the two are presented here.

Towards a more positive, behaviouristic spatial science

The main ground for disillusion within the positivist camp was a growing realization that the models being propounded and tested were not very good descriptions of reality, so that progress towards the development of geographical theory was painfully slow. Thus the large body of work based on central-place theory, for example, was built on certain axioms regarding human behaviour, with regard to choice between spatial alternatives, and from these axioms a settlement pattern was deduced. But the deductions were often only vaguely reflected in settlement morphologies, which suggested that the axioms on which they were based provided a weak foundation for understanding this aspect of the spatial organization of society. The theory suggested how the world would look under certain circumstances of economic rationality in decision-making; that those circumstances did not prevail suggested that the world should be looked at in other ways in order to understand how people do behave and structure their spatial organization. As Brookfield (1964) put it, with regard to the whole family of models then popular:

> We may thus feel that we have proceeded far enough in answering our questions when, by examination of a sufficient number of cases, we can make assertions such as the following: population density diminishes regularly away from metropolitan centres in all directions; crop yields diminish beyond a certain walking distance from the centres of habitation; air-traffic centres lying in the shadow of major centres do not command the traffic that their populations would lead us to expect. . . . Such answers, which represent the mean result of large numbers of observations whether statistically controlled or otherwise, are valuable in themselves, and sufficient for many purposes. But each is also an observation demanding explanations which may seem self-evident, or which may in fact be very elusive. Furthermore, there will be exceptions to each generalization, and in many cases there are also limits to the range of territory over which they hold true. Both the exceptions and the limits demand explanation (p. 285).

Rationality in land-use decisions

One of the first attempts by geographers to discover the 'exceptions and the limits' was the series of investigations into human responses to environmental hazards, initially floods, which was organized at the University of Chicago during the late 1950s and early 1960s. Its director was Gilbert White, whose own thesis on human adjustment to floods was published in 1945. His associates developed a behaviourist approach for studying reactions to the hazards, basing this on Herbert Simon's (1957) theories of decision-making. Thus Roder (1961), for example, categorized Topeka residents according to their attitudes to the probability of future floods there, concluding that:

> Flood danger is only one of the variables affecting the choices of the flood-plain dweller, and many considerations operate to discourage a resident from leaving the flood plain, even when he is aware of the exact hazard of remaining (p. 83).

Such behaviour, it seemed, did not fit easily into the notions of rational, profit-maximizing decision-making on which geographical theories were currently being built.

A major exponent of the behaviouristic approach was Kates (1962), who began his study of flood-plain management with the statement that 'The way men view the risks and opportunities of their uncertain environments plays a significant role in their decisions as to resource management' (p. 1). In studying such decision-making, Kates developed a schema which he claimed was relevant to a wide range of behaviours. It was based on four assumptions.

1   Men are rational when making decisions. Such an assumption may be either prescriptive—describing how men should behave—or descriptive

of actual behaviour. The latter seems to be the most fruitful, both for understanding past decisions and for predicting those yet to be made. Kates suggested adoption of Simon's concept of bounded rationality as a basis for such study; according to this, decisions are made on a rational basis, but in relation to the environment as it is perceived by the decision-maker, which may be quite different from either 'objective reality' or the world as seen by the researcher.

> In this model, men bounded by inherent computational disabilities, products of their time and place, seek to wrest from their environments those elements that might make a more satisfactory life for themselves and their fellows (p. 16).

Rational decision-making is constrained, therefore, and is not necessarily the same as the maximum rationality assumed in the neo-classical normative models discussed in earlier chapters of this book.

2   Men make choices. Many decisions are either trivial or are habitual so that they are accorded little or no thought immediately before they are made. Some major decisions regarding the environment and its use may also be habitual, but such behaviour usually only develops after a series of conscious choices has been made which leads to a stereotyped response to similar situations in the future.

3   Choices are made on the basis of knowledge. Only very rarely can a decision-maker bring together all of the information relevant to his task, and frequently he is unable to assimilate and use all that he has. 'Thus, a descriptive theory of choice must deal with the well informed and the poorly informed and the choices that men make under certainty, risk or uncertainty . . . such a theory must deal with the eventuality that not only do the conditions of knowledge vary, but the personal perception of the same information differs' (p. 19).

4   Information is evaluated according to predetermined criteria. In habitual choice the criterion is what was done before, but in conscious choice the information must be weighed according to certain rules. Some normative theory prescribes maximizing criteria (of profits, for example); descriptive theory may use Simon's notion of satisficing behaviour, involving decision-makers who seek a satisfactory outcome (a given level of profit, perhaps) only.

The results of decision-making which do not match the predictions from the sorts of theories on which Garrison, for example, based his work do not necessarily imply irrational behaviour, therefore. Almost certainly they do not. Instead, most decisions are made rationally on the basis of a, probably non-random, selection of information, are intended to satisfy some goal which is not to make the most perfect decision, and are based on criteria which vary somewhat from individual to individual. Having learned a satisfactory solution to a given class of problems, decision-makers will then continue to apply it every time such a problem occurs, unless changed circumstances require a re-evaluation.

Kates's study aimed to understand why people choose to live in areas which are prone to flooding. Their information was based on their knowledge and experience, and they could be scaled according to the certainty of their perceptions regarding further floods. In justifying their decisions, most were boundedly rational, and had made conscious choices in order to satisfy certain objectives. Similar work was reported by a number of others associated with White's leadership, and, as he points out, their findings had some impact on public policy formulation in the United States (White, 1973). Their impact on the wider geographical enterprise was not as great, however, especially in the early years of their work. This was perhaps because they were operating on the boundaries between human and physical geography, which few American researchers approached, and it was probably the later work of others which brought Simon's ideas more forcefully before the geographical audience.

Wolpert and the decision process in a spatial context
For many human geographers, it was probably a paper published by Julian Wolpert in the *Annals of the Association of American Geographers* for 1964 which introduced them to the behaviourist alternative to the normative approaches then popular. (Wolpert's paper was based on his PhD thesis submitted to the University of Wisconsin, and is further testament to the innovative qualities of the geographers there, and of their contacts with other social scientists, including agricultural economists. His research was conducted in Sweden, another hearthland of geographical innovations.) The normative theory espoused by Garrison and his followers assumes an economic man who, according to Wolpert (1964),

> is free from the multiplicity of goals and imperfect knowledge which introduce complexity into our own decision behavior. Economic Man has a single profit goal, omniscient powers of perception, reasoning, and computation, and is blessed with perfect predictive abilities . . . the outcome of his actions can be known with perfect surety (p. 537).

In the study of spatial patterns, however,

> Allowance must be made for man's finite abilities to perceive and store information, to compute optimal solutions, and to predict the outcome of future events, even if profit were his only goal (p. 537).

Thus farmers face an uncertain environment—both physical and economic—when they make their individual land-use decisions, which in aggregate comprise the land-use map. Wolpert suggested that differences between these decisions and those that would be made by 'economic man' should reflect aspects of the farmers' economic and social environments.

Comparing the labour productivity of farms in an area of Sweden with what could have been achieved under optimizing decision-making, Wolpert decided that the farmers were probably satisficers, although such a hypothesis is difficult to verify without detailed knowledge of aspiration levels. How they acted was undoubtedly contingent upon their available information, and clear spatial variations in the levels of potential productivity achieved suggested parallel spatial variations of knowledge. Only conspicuous alternatives are considered, it was suggested, and the result is rational behaviour, adapted to an uncertain environment.

Wolpert (1965) continued this theme with studies of migration, aiming to model the decision-making which lies behind the patterns of migration reported in census volumes and assiduously analysed by spatial scientists. To him, the gravity model is inadequate as a representation of such flow patterns; indeed 'Plots of migration distances defy the persistence of the most tenacious of curve fitters' (p. 159). Boundedly rational man makes decisions, first, whether to move, and secondly, where to, and he does so on the basis of his place utilities, his evaluations of the degree to which each location, including that which he currently occupies, meets his defined needs. Not only is the information on which he bases these utilities far from complete; for many places he has none. Thus each individual has an action space—'the set of place utilities which the individual perceives and to which he responds' (p. 163)—whose contents may deviate considerably from that portion of the 'real world' which it purports to represent. Once the first decision—to migrate—has been made, then the action space may be changed as the potential mover searches through it for potential satisfactory destinations and, if necessary, extends the space if no suitable solution to the search can be found.

Wolpert's papers heralded—certainly in timing and to some extent in influence too—the development of what was termed a behavioural geography (Cox and Golledge, 1969) 'united by a concern for the building of geographic theory on the basis of postulates regarding human behaviour . . . upon social and psychological mechanisms which have explicit spatial correlates and/or spatial structural implications' (pp. 1–3). This behavioural approach has focused on topics related to decision-making in spatial contexts. Golledge (1969, 1970), for example, has looked at models of learning about space and of habitual behaviour, and with Brown has investigated methods of spatial search (Golledge and Brown, 1967). Others have researched into information flows, on which decisions are based, indicating the influence of local context on behaviour (Cox, 1969), and Brown and Moore (1970) have extended Wolpert's place utility and action space concepts for the study of intra-urban migration.

The aim in behavioural geography, according to a review by Golledge, Brown and Williamson (1972), has been to derive alternative theories to those based on economic man, 'more concerned with understanding why certain activities take place rather than what patterns they produce in space' (p. 59) which involves 'the researcher using the real world from a perspective of those individuals whose decisions affect locational or distributional patterns, and . . . trying to derive sets of empirically and theoretically sound statements about individual, small group, or mass behaviours' (p. 59). In their evaluation of such behaviouristic endeavours, Golledge, Brown and Williamson indicate the seminal influence of Hägerstrand (1968), who used the concept of the mean information field (analogous to the action space of a place's residents) to model migration flows and the adoption of innovations. The initial period, 1960–65, involved the growing interest in resource-management decisions, as outlined above, and was followed by an extension of interest from environmental perception and decision-making into aspects of attitudes and motivation. These were applied to studies of migration, the diffusion of innovations, political behaviour—especially voting, perception, choice behaviour, and spatial search and learning. By studying behavioural processes in these contexts, the aspiration was to increase geographers' understanding of how spatial patterns evolve, thereby complementing their existing ability to describe such patterns. Morphological laws and systems are insufficient of themselves for understanding; the amalgamation of concepts about decision-making taken from other social sciences with geography's spatial variable would allow development of process theories that could account for the morphologies observed.

Further developments include another series of pioneering papers by Wolpert and his associates, this time relating to political decision-making. Regarding the distribution of certain artefacts in the landscape, Wolpert (1970) pointed out that the location of, for example, a public facility in an urban area frequently is the product of policy compromise:

> Sometimes the location finally chosen for a new development, or the site chosen for a relocation of an existing facility, comes out to be the site around which the least protest can be generated by those displaced. Rather than being an optimal, a rational, or even a satisfactory locational decision produced by the resolution of conflicting judgements, the decision is perhaps merely the expression of rejection by elements powerful enough to enforce their decision that another location must not be used. . . . These artefacts are rarely 'the most efficient solutions', and frequently not even satisfactory neither for those responsible for their creation nor for their users (p. 220).

(This argument avoids considering any definition of optimal, in either or both of economic and political terms.) Such decisions involve what Wolpert terms maladaptive behaviour. The work of Kates and others had suggested that decisions are adaptively rational, within the constraints of uncertainty, utility, and problem-solving ability. Coping strategies under the mutual exchange of threats between interested parties, however, can lead to decision-making which does not involve the careful and methodical investigation of alternatives until a satisfactory solution is found. Instead decisions are the consequences of conflict between groups with different attitudes and motivations, and are not the result of joint application of criteria on whose relevance there is a consensus.

> This formulation then lays the framework for the interpretation of locational decisions which appear to be more the product of stress responses than the end result of a dispassionate and considered selection of alternatives posited by the classical normative approaches or even the Simon scheme of bounded rationality (p. 224).

Wolpert and his associates have applied this formulation in a variety of contexts, such as the routes for intra-urban freeways and the siting of community mental-health facilities (Wolpert, Dear and Crawford, 1975).

Mental maps

One aspect of behavioural analysis enthusiastically adopted by a number of workers was the concept of the mental map of the environment which guides the deliberations of decision-makers. The term mental map was not new to the geographical literature, having been used by Wooldridge (1956) in his descriptions of the perceived environments within which farmers make their land-use decisions; the seminal paper which revived it was published by Gould in 1966. Gould's (1966) guiding belief was that

> If we grant that spatial behavior is our concern, then the mental images that men hold of the space around them may provide a key to some of the structures, patterns and processes of Man's work on the face of the earth (p. 182—reference to 1973 reprint).

Increasingly, he argued, location decisions are being taken with regard to perceived environmental quality, so it is necessary to know how people evaluate their environment, and whether their views are shared by their contemporaries. To investigate such questions, Gould required respondents in various countries to rank-order places according to their preferences for them as places in which to live, and these rankings were analysed to identify their common elements—the group mental maps (Gould and White, 1974). Such maps, it was argued, are useful not only in the analysis of spatial behaviour but also in the planning of social investment—such as differential salaries to attract people to less desirable areas.

Those who followed Gould's lead investigated a range of methods for identifying and analysing spatial preferences (Pocock and Hudson, 1978). The result obtained provided little input to theoretical development, however, and Downs (1970) wrote that

Even the most fervent proponent of the current view (that human spatial behaviour patterns can partially be explained by a study of perception) would admit that the resultant investigations have not *yet* made a significant contribution to the development of geographic theory (p. 67).

Apart from Gould's rank-ordering procedures, Downs identified two other major approaches to the study of environmental images: the structural approach, which inquires into the nature of the spatial information stored in people's minds and which they use in their everyday lives— Lynch's (1960) book was a model for such work—and the evaluative approach, in which (Downs, 1970) 'The question is, what factors do people consider important about their environment, and how, having estimated the relative importance of these factors, do they employ them in their decision-making activities' (p. 80). With this evaluative approach geographers moved into the wider field of cognitive mapping—'a construct which encompasses those cognitive processes which enable people to acquire, code, store, recall and manipulate information about the nature of their spatial environment' (Downs and Stea, 1973, p. xiv). In this they worked alongside both psychologists, who were becoming increasingly interested in man's relationship to a wider area than his proximate environment and in the development of relevant non-experimental research techniques, and designers concerned with the creation of more 'liveable' environments. The journal *Environment and Behavior* was launched in 1969 to cater for this interdisciplinary market, but despite some interest (e.g. Tuan, 1975a; Downs and Stea, 1977; Pocock and Hudson, 1978) the general field has not made a major impact within human geography.

Time-space geography
During the late 1960s, Hägerstrand began to formulate a framework for the study of human geography based on a joint time-space set of coordinates; this work was only publicized outside Sweden in the early 1970s (Pred, 1973). As interpreted to a wider audience (Carlstein et al., 1978), time and space are viewed as resources which constrain activity. Any behaviour which requires movement involves the individual or group traversing a path through space and time, as depicted in Figure 5.1 in which movements along the horizontal axis indicate spatial traverses while those along the vertical mark the passage of time. Any journey, termed a lifeline by Hägerstrand, thus involves movement along both axes simultaneously.

Constraints to movement in time and space operate in three ways. First, there are capability constraints restricting movement; in time, these refer to the biological need for about eight hours of sleep in every twenty-four, whereas the extent of movement in space during a given time is controlled by the available means of transport. Secondly, there are coupling constraints, which operate to require certain individuals

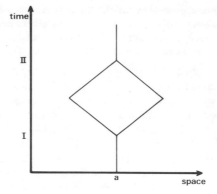

Figure 5.1   The time-space prism. In this simple example a person starts at point a: he cannot leave this point until time I and he must return to it by time II: the prism between those times indicates his maximum available spatial range.

and groups to be in particular places at stated times (teachers and pupils in schools, for example). And finally, authority constraints preclude an individual from being in certain places at set times. Together, these three define the time-space prism (Figure 5.1) which contains all the available lifeline paths for an individual starting at a certain location and needing to return there at a given time.

Pred (1977) has claimed of Hägerstrand's time-space geography that it

> has the potential for shedding new light on some of the very different kinds of questions customarily posed by 'old-fashioned' regional and historical geographers, as well as 'modern' human geographers . . . [it] holds the promise of identifying new questions of social and scholarly significance. It may also satisfy the longing shared by many for a more humanistically oriented geographic approach to some of the more complex and frustrating problems of modern society . . . it has the potential for enabling geographers to attain a new level of intellectual maturity. That is, through Hägerstrand's conceptual structure geographers may grow so bold as to seriously ask themselves what insights and assistance they can provide to researchers in the neighbouring social and biological sciences (p. 207).

Hägerstrand's intention with his framework, sometimes referred to as the study of chorography, is to provide a focus on the quality-of-life implications of the packing of people in time and space, and instead of simply looking, as in traditional geographical investigations, at parts of a structure it offers a contextual approach which views individuals' situations relative to other individuals within their environments. A new regional geography may emerge emphasizing 'principles of togetherness' (Pred, 1977, p. 213) rather than landscape components, and planning applications have been developed for the siting of facilities so that they are accessible within the constraints of time-space prisms (Carlstein,

Parkes and Thrift, 1978; Pred and Palm, 1978). Thus, according to Pred (1977) although

> Time-geography is not a panacea for human geographers . . . It is, however, a great challenge . . . to cease taking distance itself so seriously . . . to accept that space and time are universally and inseparably wed to one another; to realize that questions pertaining to human organization of the earth's surface, human ecology, and landscape evolution cannot divorce the finitudes of space and time. . . . It is a challenge to turn to the 'choreography' of individual and collective existence—to reject the excesses of inter- and intra-disciplinary specialization for a concern with collateral processes (p. 218).

Methods in behavioural geography

Whereas the theory- and model-builders of the spatial-science school of human geographers received much of their stimulus from neo-classical economics, in some cases via regional science, the alliance for behavioural geography was largely with the social sciences having a larger empirical content, notably psychology and sociology. The behaviouristic approach is an inductive one, with the aim being to build general statements out of observations of ongoing processes. The areas studied were very much determined by the work in the spatial-science school. As Brookfield's statement quoted at the beginning of this chapter suggests, the models and theories of the latter, such as central-place theory, raised many of the queries which the behaviourists, stimulated by their observations of the failings of such theories when matched against the 'real world', sought to follow-up. In terms of the accepted route to scientific explanation (Figure 3.1), therefore, behavioural geography involved moving outside the accepted cyclic procedure to input new sets of observations on which superior theories might be based. In doing this, the behaviourists did not really move far from the spatial-science ethos. Indeed, many of their methods were those of their predecessors; Gould's mental-map studies, for example, used the same technical apparatus as the factorial ecologies (p. 73).

Somewhat away from this general orientation of behavioural work, Pred (1967, 1969) presented an ambitious alternative to theory-building based on 'economic men' in his two-volume work *Behavior and Location*. He began with a critique of existing location theory comprising three groups of objections: those concerned with logical inconsistency—it is impossible for competing decision-makers to arrive at optimal location decisions simultaneously; those concerned with motives—maximizing versus satisficing; and those concerned with human ability to collect, assimilate and use all possible information. Thus (Pred, 1967):

> Bunge's theoretical geography is easily distinguished from geographical location theory because its optimal final goals are disassociated from the interpretation of real-world economic phenomena . . . and because these same goals can only yield a body of theory that for all intents and purposes is totally abehavioral and static rather than dynamic (p. 17).

Decisions on locations and on land use are made with imperfect knowledge by fallible individuals. As a result there is bound to be some
disorder in the ensuing spatial patterns.

As an alternative to the forms of theory-building which he attacked,
Pred proposed the use of a behavioural matrix (Figure 5.2), whose axes
are quantity and quality of information available and the ability to use
that information: economic man is located in the bottom right-hand
corner. Because of the nature and importance of information flows,
position on the first axis depends in part on the decision-maker's spatial
location; on the second, his position would reflect aspiration levels,
experience, and the norms of any groups to which he belonged. Different
people in different positions in the matrix would vary in their decisions,
therefore; indeed, two at the same position may act on different bases and
in different ways.

Individuals do not stay at the same position in the matrix, and spatial
patterns are not static. Thus in his second volume, Pred (1969) introduced a dynamic element by shifting individuals through the matrix; as
they shift, and change their decisions, so the environment changes for
others. As people learn, they both acquire more and better information,
and become more skilled in its use; they shift towards the bottom
right-hand corner of the matrix, some of them in advance of others, who
benefit from the activities of the 'decision-leaders'. The unsuccessful are
gradually eliminated, so that with time a concentration of 'good'
decision-makers close to the economic-man position evolves. But
changes in the external environment produce parametric shocks, which
result in decision-makers becoming less informed and less certain; as a
consequence they are shifted back towards the upper left-hand corner
and another learning cycle begins. As long as parametric shocks occur

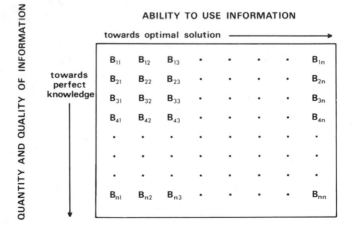

Figure 5.2   The behavioural matrix. Source: Pred (1967, p. 25).

more frequently than the learning experience takes, an optimal location pattern will never emerge, except perhaps by chance.

Pred (1969) presented the behavioural matrix as a 'gross first approximation' (p. 91), arguing that any theory is better than none (p. 139), even if the model itself is literally untestable (p. 141). Harvey (1969b) treated it to a scathing review, however, calling the two dimensions of the matrix vaguely defined, ambiguous, unoperational, and an oversimplification of the complex nature of behaviour. Indeed, Harvey (1969c) was generally sceptical about the potential of a behavioural location theory, a view shared by Olsson (1969) who pointed out the difficulties of studying processes and demonstrated that much behavioural geography involved only the inferring of processes from aggregated data on individual behaviour. Others argued that such inference could be very strong, however. Rushton (1969), for example, accepted that any one pattern of behaviour—what he termed *behaviour in space*—was largely a function of the spatial structure within which it occurred (the choice of shopping centres by a population, for example), but claimed that general rules of *spatial behaviour* could be deduced from examining the types of preferences displayed within a particular pattern:

> To say that these preferences do not exist independently of the environment where the decision is made is to argue that environments could exist about which the person would be unable to reach a decision (p. 393).

Thus a distance-decay pattern reflects the details of the environment in which it is observed, but to produce it people must make decisions based on certain rules which they apply, and Rushton developed a procedure for isolating those rules. Its validity has been queried, however (Pirie, 1976), and, like Pred's behavioural matrix, it has not attracted many disciples.

Harvey (1969c) suggested two alternatives to behavioural location theory—further development of normative theory and the construction of a stochastic location theory. He felt that either of these offered more immediate pay-offs in terms of understanding spatial patterns than did behavioural efforts, because of the conceptual and measurement problems of the latter. The stochastic approach was also favoured by Curry (1967b), who argued that large-scale patterns are the outcomes of small-scale indeterminacy; individual choices may be random within certain constraints, but when very many of them are aggregated they may display considerable order (see above, p.108). Similarly, Webber (1972) attempted to model locational decision-making processes in states of uncertainty using normative approaches: his conclusion was that 'uncertainty increases agglomeration economies' (p. 279), thereby leading to greater concentration of economic activities and people into cities than would be predicted by models based on 'economic men'. Game theory is a mathematical procedure developed to handle decision-making in uncertainty, but it has received little attention from human geographers (Gould, 1963). The most ambitious attempt to build on such theories, which has had very little impact, is Isard *et al.* (1969).

The behavioural approach has not brought about a revolution away from the spatial-science paradigm, therefore, and in effect has become an attachment to it. Whereas various normative approaches start with certain assumptions, usually simplifying ones, about human behaviour, and then deduce what spatial patterns follow from such axioms, the behaviourists have sought to modify the assumptions by an inductive procedure which seeks the rules of behaviour that can be used to predict (and therefore explain) spatial patterns. In its entirety, this latter approach involves a sequence of interrelated investigations. An individual is faced with a decision, either one with a direct spatial input or one with spatial consequences. To make that decision, he sets criteria, collects information, and then evaluates the information against the criteria. As a result of the evaluation, a decision to act might be made, or instead he determines to change his criteria, to collect more information, or to do both. (In this way, his problem-solving is very similar to the procedures of the scientific method outlined in Figure 3.1.) Many investigations have not gone through the full sequence, however, but have focused on certain aspects of it, such as the flow of information, from which characteristics of other elements in the sequence may be inferred.

Whereas very many of the studies in the spatial-science school reviewed in earlier chapters could be conducted using either published data sources (such as censuses) or relatively small field-collection exercises, the behavioural approach has required much more effort on the part of human geographers in data collection from the individual decision-makers. This need to conduct social surveys of various kinds has furthered the growing links between geographers and sociologists, psychologists, and, to a lesser extent, political scientists, and has led to an expansion in the data-handling procedures necessary for the training of geographers. One of the problems in their use is that many of the topics studied in human geography involve very large numbers of individual decisions—as in migrations, journeys to work and to shop, voting decisions, and so on. Very large sample surveys may be necessary to produce valid generalizations about such behaviour, but resource limitations have meant concentration on both small selections and only limited segments of the full behavioural sequence. Thus, for example, Brown and Moore's (1970) schema for the study of intra-urban-migration decisions has mainly been tested in part only (e.g. Clark, 1975). More success with use of the behavioural sequence has probably been achieved in those branches of human geography that deal with topics involving relatively few decision-makers. In the study of diffusion, for example, Brown (1975) has attempted to divert attention away from overall patterns of spatial spread and the reasons for adoption (or not) to the decision-making which brings certain innovations to places; most of these innovations involve the selling of products, and so he terms them 'consumer innovations'. Similarly, a number of industrial geographers have moved from investigations of aggregate patterns, which could be compared to those predicted by application of neo-classical economic analysis (Smith, 1971), to the study of decision-making behaviour by firms (e.g. Hamilton, 1974).

Cultural and historical, behavioural geography

As indicated in earlier chapters (e.g. pp. 70 and 79), workers in cultural and historical geography were neither closely involved in nor attracted by the so-called quantitative and theoretical revolutions. There was some application of statistical procedures, as attempts to make work in these fields appear more 'modern'—several of these attempts were by non-historical (e.g. Pitts, 1965) and non-cultural geographers—but few changes of substance in the approaches adopted. But by the 1970s, historical and cultural geographers had taken the initiative, and were proposing alternative methodologies to that of positivism, methodologies which were humanistic in their orientation.

The beginnings of these changes can be traced to two papers, one of which had more impact than the other on the geographical discipline at large. In the first, John K. Wright (1947) introduced the term *geosophy*, defined as the study of geographical knowledge:

> it covers the geographical ideas, both true and false, of all manners of people—not only geographers, but farmers and fishermen, business executives and poets, novelists and painters, Bedouins and Hottentots—and for this reason it necessarily has to do with subjective conceptions (p. 12).

Wright conceded that study of such subjective ideas was not open to the employment of the strict scientific principles of physical geography, but claimed that it provided indispensable background and perspective to geographical work:

> geographical knowledge of one kind or another is universal among men, and is in no sense a monopoly of geographers . . . such knowledge is acquired in the first instance through observations of many kinds. . . . Its acquisition, in turn, is conditioned by the complex interplay of cultural and psychological factors . . . nearly every important activity in which man engages, from hoeing in a field or writing a book or conducting a business to spreading a gospel or waging a war, is to some extent affected by the geographical knowledge at his disposal (pp. 13–14).

These words could well have heralded an earlier start to behavioural geography than that chronicled earlier in this chapter, but apparently, by the lack of reference to them in later published works, they had little impact until taken up by Wright's colleague at the American Geographical

Society, David Lowenthal (1961), in a widely cited paper 'concerned with *all* geographic thought, scientific and other: how it is acquired, transmitted, altered and integrated into conceptual systems' (p. 259). In a wide-ranging survey, Lowenthal argues that the world of each individual's experience is intensely parochial and covers but a small fraction of the total available. There are consensus views about many aspects of the world, but often an individual will mistakenly assume that his view is the consensus. We all live in personal worlds, which are 'both more and less inclusive than the common realm' (p. 248). Our perceptions of these personal worlds are personal too; they are not fantasies, being firmly rooted in reality, but because 'we elect to see certain aspects of the world and to avoid others' (p. 251) behaviour based on such perceptions must have its unique elements. Different cultures have their own shared stereotypes, however, which are often reflected in language, and they attempt to create environments fitting into these stereotypes:

> The surface of the earth is shaped for each person by refraction through cultural and personal lenses of custom and fancy. We are all artists and landscape architects, creating order and organizing space, time, and causality in accordance with our apperceptions and predilections (p. 260).

These ideas were put into practice in other papers concerned with the interpretation of landscapes as reflections of societal norms and tastes (e.g. Lowenthal, 1968; Lowenthal and Prince, 1965), thereby belatedly bringing Wright's ideas before a wider, and more readily appreciative perhaps, audience.

The second of the original papers was by a British geographer, although it was published in India (Kirk, 1951): the main arguments were reiterated in a later article (Kirk, 1963). He, too, stressed that the environment is not simply a 'thing' but rather a whole with 'shape, cohesiveness and meaning added to it by the act of human perception' (Kirk, 1963, p. 365): once this meaning has been ascribed, it tends to be passed to later generations. Thus Kirk recognizes two, separate but not independent, environments; a phenomenal environment, which is the totality of the earth's surface, and a behavioural environment, which is the perceived phenomenal environment:

> Facts which exist in the Phenomenal Environment but do not enter the Behavioural Environment of a society have no relevance to rational, spatial behaviour and consequently do not enter into problems of the Geographical Environment (p. 367).

Since much of geography is concerned with decision-making and its consequences, appreciation of the behavioural environment should be central to the study of geography. Indeed, according to Ley (1977), one cannot proceed without such awareness of what is in the behavioural environment; even an apparently neutral statement such as 'Pittsburgh is a steel town' is, he argues, a value-laden view of a geographer-outsider, which may not accord with the perceptions of the resident-insiders. Thus:

> Too often there is the danger that our geography reflects our own concerns, and not the meanings of the people and places we write of. . . . The geographical fact is as thoroughly a social product as the landscape to which it is attached (p. 12).

Scientific disciplines, as indicated in Chapter 1, can be divided into their 'invisible colleges', groups of scholars working on the same topic who refer to each other's publications. Wright, Lowenthal and Kirk were not members of any major college during the early 1960s, and so had very little impact on the first phase of behavioural work identified earlier in this chapter. (None of the three, for example, is in Pred's (1967, 1969) bibliographies, nor is any one referred to by Wolpert: Golledge, Brown and Williamson (1972, p. 75) make only a passing reference to Lowenthal's work—'Pursued by insightful researchers, the analysis of literary and other artistic data of past and present can have strong explanatory power. The subjective element in these attempts to assess the impact of spatial perception is acknowledged, but its presence in many other studies is more subtle and potentially damaging'.) The behavioural work of the mid-1960s was in the positivist mould; that of the three authors just discussed was not.

Several cultural and historical geographers took up the concepts of the behavioural or perceived environments. Among the former, one of the leaders was Brookfield, a British geographer with field experience, by the mid-1960s, of South Africa, Mauritius, New Guinea, and the Pacific Islands. In reviewing work by cultural geographers on alien societies, he noted (Brookfield, 1964):

> A difference of approach is apparent between those who have an overtly chorographic purpose, who scarcely ever seek explanations in matters such as human behaviour, attitudes and beliefs, social organization, and the characteristics and interrelationships of human groups, and those whose inquiries are not primarily chorographic, and who are more inclined to undertake a search for processes as a means of reaching explanation (p. 283).

Social organization, Brookfield argued, is the key to many explanations, so that:

> when an individual human geographer is sitting down in one small corner of a foreign land, and seeks to interpret the geography of that small corner, then it is difficult for him to do so without trying to comprehend the perception of environment among the inhabitants (p. 287).

But geographers had largely failed to delve into such details of social organization, because of the broad areal scale at which they had tended to work, their concern with distributions rather than with processes, and their avoidance of what he terms 'micro-geography'. Inquiry in human geography should involve three stages:

1   general statements about areal patterns and interrelationships;
2   detailed local inquiries which follow up the questions about processes raised by these general statements; and
3   organization of the general and local material to produce explanatory generalizations.

Brookfield's argument was for more micro-geographical studies at the second of these stages, providing the basis for the development of comparative methods with which generalizations could be forged (Brookfield, 1962).

Brookfield (1969) later surveyed the literature which showed that 'decision-makers operating on an environment base their decisions on the environment as they perceive it, not as it is. The action resulting from decision, on the other hand, is played out in a real environment' (p. 53). Referring to the 'modern' behavioural work, as well as the studies of cultural geographers, he pointed out the great problems involved in isolating the perceived environment—something which is 'complex, monistic, distorted and discontinuous, unstable and full of unwoven irrelevancies' (p. 74)—and in building it into an analytical methodology. Further data are needed, too, on such topics as work-organization, time-allocation and budget-allocation, on the meaning of consumption and of distance—all necessary tasks for the full understanding of man-environment systems.

The concept of the perceived environment has a considerable pedigree in historical-geographical scholarship, though without the current terminology; a major example of its use is Glacken's (1956, 1967) survey of societal attitudes to environments. Historical geographers, it has been suggested (Prince, 1971a), must study a trilogy of worlds: the real world, as recorded in documents and in the landscape; the abstract world, as depicted by general models of spatial order in the past; and the perceived world: 'Past worlds, seen through the eyes of contemporaries, perceived according to their culturally acquired preferences and prejudices, shaped in the images of their assumed worlds' (p. 4). From these three, it may be possible to provide explanations of landscape changes, which cannot be obtained from the assumed processes provided by continua of data over time (see also Moodie and Lehr, 1976). Reconstruction of past environments is extremely difficult, however, for it involves seeing the written record through the cultural lens of the writer:

> A study of past behavioural environments provides a key to understanding past actions, explaining why changes were made in the landscape. We must understand man and his cultures before we can understand landscapes; we must understand what limits of physical and mental strain his body will bear; we must learn what choices his culture makes available to him and what sanctions his fellows impose upon him to deter him from transgressing and to encourage him to conform (Prince, 1971a, p. 44).

Perhaps the enormity of such a task is the reason why most success in such reconstructions has been with regard to the relatively recent past, such as of the perceptions which guided the settlement of the American West (e.g. Lewis, 1966), although Wright (1925) essayed a similar task for Europe at the time of the Crusades.

Not all perceived worlds refer to either past or present; some landscapes have been fashioned by men out of their Utopian views of the future (Porter and Lukermann, 1975; Powell, 1971). In general, however, and whether of past, present or future, geosophy has not become a popular field of study, despite some intriguing essays (see Lowenthal and Bowden, 1975). But directly or indirectly, it has led to arguments for alternative approaches in human geography to those of the positivist, and it is those arguments which form the material for the rest of this chapter.

The attack on positivism and the hermeneutic approaches

From the early 1970s, work by cultural and historical geographers has been presented as an attack on the positivism of spatial science. To replace the latter, a variety of hermeneutic approaches have been proposed, presenting a focus on the decision-maker and his perceived world and denying the existence of an objective world which can be studied by positivist methods. The intent is to reorient human geography towards a more humanistic stance, to resurrect its synthetic character, and to re-emphasize the importance of studying unique events rather than the spuriously general.

Anti-positivism, idealism, and historical geography

The various hermeneutic (or interpretative) approaches have much in common, but they can be separated into different proposals in the present context. The first to be discussed here is associated with two workers who were together at the University of Toronto at the end of the 1960s; both are historical geographers.

The basic theme of a first paper, published by the senior of these two authors (Harris, 1971), is that geography is a synthetic discipline, concerned with particular assemblages of phenomena and not with the science of spatial relations. Thus:

> When the history of North American geography in the 1950s and 1960s is written, a paradox with which it will have to deal is how, with little argued, logical justification, so many geographers came to see their subject as a science of spatial relations (p. 157).

With May and Sack (p. 96), Harris sees the spatial perspective leading to the dismemberment of geography, as specialists communicate more with their contemporaries in other disciplines than with geographers, and develop theories which are descriptions of how the world might operate under certain conditions, rather than of how it does actually work.

> The difficulty in conceiving of geographical theory comes down to this. The development of theory is necessarily an exercise in abstraction and simplification in which the complexities of particular situations are eliminated to the point which common characteristics become apparent. But if geography is thought to have a particular subject matter, it is certainly not individual phenomena or categories of phenomena which other fields do not study. Rather it is a whole complex of phenomena, many or all of which may be studied individually by other fields but which are not studied elsewhere in their complex interactions (p. 162).

The clear parallel, according to Harris, is with history, for:

> Few historians would attempt to develop a general theory of revolutions. In so doing they would lose grasp of the type of insight that characterizes good historical synthesis (p. 163).

The goal of both history and geography is synthesis, therefore. In developing syntheses, positivist methods may be applicable. Historians could be law-consumers, applying the generalizations of other social scientists to particular events, and geographers could operate likewise; alternatively, both historians and geographers could apply the idealist method, arguing that all activity is based on personal theories. Thus Harris (1978) writes elsewhere of a 'historical mind':

> Such a mind is contextual, not law-finding. Sometimes it is thought of as law-applying but, characteristically, the historical mind is dubious that there are overarching laws to explain the general patterns of human life (p. 126).

This mind, he argues, is open and eclectic, uses no formal research procedures, sees things in context, is sensitive to motives and values, excludes little, and is wary of sweeping generalizations. Its goal is understanding, not planning, and this should be the case with the 'geographical mind' too.

Developing his theme of the parallel between history and geography, Harris (1971, p. 167) suggests four major points of agreement concerning the nature of history:
1 Its primary concern is with the particular;
2 Explanation may take into account the thoughts of relevant individuals;
3 Explanation may make use of general laws; and
4 'Explanation in history relies heavily on the reflective judgement of individual historians'.
From these, he then argues:

> If geography aims to describe and explain not so much particular events or peoples, as particular parts of the surface of the earth, then these points of agreement about history also apply to geography (p. 167).

(The term 'particular parts' can be widely interpreted, it would seem, for Harris (1977) has himself sought to understand the nature of north-western European colonizing societies 'by a model'.) The landscape results from actions; behind those actions lie thoughts; study of thoughts allows understanding of landscape. Thus synthesis is crucial, since:

> the idea of synthesis itself becomes more important as it becomes obvious that our larger problems transcend narrow subject-matter fields . . . integration . . . in a larger understanding is still achieved, however aided by statistical methods and computers, by the judgement of wise men who have cultivated the habit of seeing things together (p. 170).

Geographers, presumably, are to be those wise men, not an original claim for, according to Buttimer (1978a), the basis of Paul Vidal de la Blache's work was that:

> The task which no other discipline with the possible exception of history claims is to examine how diverse phenomena and forces interweave and connect with the finite horizons of particular settings. Temporality and spatiality are universal features of life so historical and geographical study belong together (p. 73).

Many of Harris's arguments were extended by his Toronto colleague, Leonard Guelke, whose first paper (Guelke, 1971) was a strong criticism of the 'narrowly conceived scientific approach' (p. 38) of geography using the positivist method. Thus he argued against geography as a law-seeking activity by asking the proponents of the positivist approach to indicate how their laws would meet the basic standards of scientific acceptability, particularly with regard to prediction. Whereas they might be able to produce generalizations concerning the phenomena which they actually studied he felt it very unlikely that they could define laws applicable to all examples of the relevant phenomena. Statistical regularities are not laws and

> Until the new geographers have shown that the laws that might conceivably be discovered in geography will be more than generalizations, which describe common but non-essential connections between phenomena, their claims must be treated cautiously . . . there is little cause for optimism, especially as the statistical methods widely employed by geographers cannot be considered appropriate law-finding procedures (p. 42).

Regarding geography as a law-applying science, Guelke argues that laws of human behaviour are virtually impossible to conceive in anything but the most generalized form, because so much behaviour is culturally-specific, and an apriori statement of the determining conditions for their operation is not feasible. Thus 'Human geographers cannot consider themselves to be law-applying scientists . . . because they have no laws to apply' (p. 45).

Turning to the use of theories and models in geography, Guelke points out that for them to serve a valid purpose in the pursuit of understanding, criteria must be erected which indicate how such devices are testable against reality (see also Newman, 1973, on the vague use of the term hypothesis). Such criteria have not been, and cannot be, stated, Guelke claims; to him, studies purporting to test central-place theory seem to operate the rule that 'one counts one's hits but not one's misses' (p. 48; see also Guelke, 1978, p. 50). Too often, failure to produce reality is explained by claims that the test environment was not entirely suitable, and frequently *ad hoc* hypotheses are adduced to account for observed disparities. Models and theories may have heuristic value for human geographers, clarifying certain aspects, therefore, but they can have no explanatory power.

Guelke's (1971) conclusions are that:

> The new geography . . . has not yet produced any scientific laws and . . . appears unlikely to produce them in the future. . . . The theories and models . . . are not amenable to empirical testing. . . . The new geographers insisted on . . . logical and internally consistent theories and models. Yet, none of their theoretical constructs were ever complex enough to describe the real world accurately. They had achieved internal consistency while losing their grip on reality (pp. 50–1).

His alternative to the so discredited procedure is the idealist approach mentioned by Harris, which is (Guelke, 1974) 'a method by which one can rethink the thoughts of those whose actions he seeks to explain' (p. 193). All actions, according to the idealist, are the result of rational thought, the parameters of which are constrained by a theory, which in turn is 'any system of ideas that man has invented, imposed, or elicited from the raw data of sensation that make connections between the phenomena of the external world' (p. 194). Many such theories are part of the society and culture inhabited by the actor under consideration, and include its religions, myths and traditions. Using them, 'the explanation of an action is complete when the agent's goal and theoretical understanding of his situation have been discovered. . . . One must discover what he believed, not why he believed it' (p. 197). Thus human geographers do not need to develop theories, since the relevant theories, which led to the action being studied, already exist (or existed) in the minds of the actors. The task of the analyst is to isolate those theories. Some of them may be unique to particular individuals—such as that which led Columbus to sail westwards—but there are many consensus theories, shared in large part by large numbers of actors; they represent the order which man has himself stamped on the world, and do not require further theories in order to be understood.

Guelke's argument was challenged by Chappell (1975), who pointed out that by focusing on the individual actor alone the idealist omitted any reference to the environmental constraints and influences on his actions (see also Gregory, 1978a). Guelke (1976) accepted the existence of such constraints and influences, but claimed that investigation of them lay outside the geographer's domain. Study of environmental causes would, he felt, lead into physiology and psychology and deviate attention from 'the most critical dimension of human behaviour, namely the thought behind it' (p. 169). Chappell (1976) responded that 'to go so far as to say that there is no possible respectable theory to explain man's rational theories and the actions which flow from them' (p. 170) is to be myopic: 'paradigms not only explain facts but they guide the research of whole disciplines' (p. 171). To him, Guelke's contention that the ultimate causes of man's actions lie outside the scope of human geography places geographers in an inferior position in the academic division of labour.

In a further essay, Guelke (1975) addressed his ideas on idealist approaches to historical geographers, as a counter to the arguments that they should adopt the approaches and techniques of positivism (see p. 79). He argued:

> It is obvious that quantitative techniques will often be useful. . . . Statistical methods put in harness with positivist philosophy are a dangerous combination. . . . Historical geographers need to rethink not their techniques but their philosophy. . . . This can best be achieved by moving from problem-solving contemporary applied geography towards the idealist approach widely adopted by historians (p. 138).

Gregory (1976) agreed with the first part of this statement, but not with the proposed solution. Like Chappell, he saw the need to investigate individual action within its constraining structures (see below, p. 163).

Logical-positivist approaches in human geography have been defended against the idealist attack by Hay (1979), who both responds to the criticisms and in return raises points of contention in the proposed alternative. Thus he argues that Guelke's case is ill-founded and rests on misconceptions of the nature of positivism, such as that all theory must be both normative and based on conceptions of economic man, that to be scientific is to be nomological, and that prediction is the same as prophesy (rather than simply testing, from the known to the unknown). Further, he claims that Guelke presents an anti-positivist argument by using a positivist test, and that he fails to realize the value of *ad hoc* hypotheses in the improvement of theory: Guelke should not, according to Hay, ask 'does this theory explain Y?' but rather should ask 'does this theory contribute to an understanding of Y?'.

With regard to the idealist alternative, Hay raises the problem of studying groups rather than individuals. To Guelke (1978):

> The assumption that thought lies behind human action is not related to the numbers involved. . . . If thousands of people drive motor cars to their places of work the idealist assumes that each of these journeys is a considered action involving thought. In such situations the investigator will not be able to look at each case individually, but he will seek to isolate the general factors involved in typical circumstances . . . [for which he] might well make use of

statistical procedures . . . the value of statistical analysis will largely depend
on its successful integration in the general interpretation or explanatory thesis
being developed (p. 55),

which is a procedure akin to that employed by the behaviourist geographers whose work has been reviewed earlier in this chapter. Such a
procedure does not give ontological status to groups as collections of
individuals in which, as with 'traditional' regions, the whole is greater
than the sum of the parts. Secondly, Hay points out that objective facts
must influence behavioural outcomes, in addition to the thoughts of the
actors: Columbus found America, because it was there. Thirdly, he
claims that the idealist position ignores the possibility of either uncon-
scious or subconscious behaviour. In sum, idealism is reductionist, but
the world is more than a large number of independent decision-makers.

Idealism has also been criticized by Mabogunje (1977) who claims that
'Such a retreat from objective theory formulations as a means of seeking
explanation to certain events would exclude from our consideration the
exploration of the consequences of societal actions' (p. 368) and that
instead of retreating to a focus of particular cases—'seek[ing] special
explanation for each situation in which a different value system can be
shown to be operative' (p. 370)—geographers should attempt to build
better theories encompassing these differences in value-systems. Others
have asked how an idealist interpretation can be verified. (See also the
same question regarding phenomenology raised by Mercer and Powell
(1972), who wonder whether two phenomenologists can ever have the
same 'intuitions' of a phenomenon, or indeed know whether they have.)
Guelke is prepared for this argument, presenting the analogy of the
court-of-law, rather than the positivist's laboratory, for his Popperian
procedure.

A well verified idealist explanation will be one in which a 'pattern of
behaviour can be shown to be consistent with certain underlying ideas.
Where data are presented which are not in accordance with a proposed
explanation a new hypothesis will be needed' (Guelke, 1978, p. 55),
which is somewhat in contradiction to his earlier statements that theory
should not be imposed from without by the observer (p. 132). Even so,
he says, 'one cannot guarantee mistake-free interpretations. The complex nature of human societies and lack of pertinent data makes it
inevitable that many idealist interpretations will be of a tentative character' (p. 55), a position which is very similar to Moss's (1977) outline of
deductive procedures in historical explanation.

Phenomenology and other humanistic approaches
Phenomenology is a hermeneutic approach which has attracted more
attention among human geographers than has idealism. The term was
first used by a geographer as long ago as 1925 (by Sauer) but it has only
become widely known during the 1970s.

The first direct statement advocating a phenomenological approach was by Relph (1970), who has also been associated with the Department of Geography at the University of Toronto. Despite a variety of specific interpretations, he noted that the basic aim of phenomenology is to present an alternative methodology to the hypothesis-testing and theory-building of positivism, an alternative grounded in man's lived world of experience. Phenomenologists argue that there is no objective world independent of man's existence—'all knowledge proceeds from the world of experience and cannot be independent of that world' (p. 193). The crucial elements of the world of experience are the *essences*— 'the elements and notions which characterize the nature of an entity or phenomenon' (p. 194)—and it is description of these essences and their role in man's consciousness which characterizes the phenomenological method. Thus, according to Entrikin (1976) 'phenomenologists describe, rather than explain, in that explanation is viewed as a construction and hence antithetical to the phenomenologist's attempt to "get back" to the meaning of the data of consciousness' (p. 617).

The essences studied by phenomenologists are related to man's intentionality as a rational actor, since, as the idealists also argue, human behaviour can only be understood in the context of human thought. Phenomenology focuses on that thought, on the world as man experiences it. Thus resources, for example, are human appraisals, not objective facts existing outside the lived world, and such appraisals are a mixture of consensual and individual views. Adoption of the phenomenological method by geographers allows for unified study of man in his environment.

> If geography is thought to be concerned primarily with the development of 'objective' laws and theories, the criticisms of phenomenology, at least, should not be ignored. But if geography is thought to be concerned in some way with understanding man on the human level, then the concepts and methods of phenomenology have much to offer (Relph, 1970, p. 199).

Relph's paper was followed by another from a geographer having associations with the University of Toronto, Yi-Fu Tuan (1971), to whom geography is the mirror of man, revealing the essence of human existence and striving: to know the world is to know oneself, just as careful analysis of a house reveals much about both the designer and the occupant. Thus the study of landscapes is the study of the essences in the societies which mould them, in just the same way that the study of literature and art reveals much of human life. Such study, by geographers, has a clear base in the humanities, rather than the social or physical sciences; Tuan (1974, 1975b) has illustrated it in a number of essays, giving, for example, insights into such topics as the sense of place.

> Humanistic geography achieves an understanding of the human world by studying people's relations with nature, their geographical behaviour as well

as their feelings and ideas in regard to space and place (Tuan, 1976, p. 266). Scientific approaches to the study of man tend to minimize the role of human awareness and knowledge. Humanistic geography, by contrast, specifically tries to understand how geographical activities and phenomena reveal the quality of human awareness (p. 267).

Tuan exemplifies this with five themes: the nature of geographical knowledge and its role in human survival; the role of territory in human behaviour and the creation of place identities; the interrelationships between crowding and privacy, as mediated by culture; the role of knowledge as an influence on livelihood; and the influence of religion on human activity. Such concerns are best developed in historical and in regional geography; their value to human welfare is that they clarify the nature of the experience (see also Appleton, 1975). Indeed, Tuan (1978) claims that 'The model for the regional geographers of humanist leaning is . . . the Victorian novelist who strives to achieve a synthesis of the subjective and the objective' (p. 204). He quotes the first two pages of E. M. Forster's *A Passage to India* as a paradigm example, and Glacken (1978) presents a similar case, citing *The Road to Xanadu*.

Among others advocating the phenomenological approach, Mercer and Powell (1972) echo the arguments that application of positivism in geography 'left the subject with too many technicians and a dearth of scholars' (p. 28). Land-use patterns, they claimed, can never be understood 'by the elementary dictates of geometry and cash register' (p. 42); the world can only be comprehended through man's intentions and his attitudes towards it. In a lengthy discussion of the nature of phenomenology and its development in other disciplines, notably sociology, they point out 'a very real danger of the research-worker assuming that concepts which are cognitively organized in his own mind "exist" and are equally clearly organized in the minds of his respondents' (p. 26), and argue instead for research methods which lead to empathy between observer and observed. Within geography this requires 'that we make every effort to view problems and situations not from our own perspective, but from the actor's frame of reference' (p. 48)—which is a scientific position of 'disciplined naivete'.

In similar vein, Anne Buttimer made a general case for study by geographers of the values which permeate all aspects of living and thinking (Buttimer, 1974; see also Glacken, 1978). She feels that the order, precision and theory developed by positivist social sciences are dearly bought—'we often lose in adequacy to deal with the values and meanings of the everyday world' (p. 3)—and that the behavioural geography of the sort discussed earlier in this chapter does not break away from the mechanistic, natural-science view of man. On the other hand 'An existentially aware geographer is . . . less interested in establishing intellectual control over man through preconceived analytical models than he is in encountering people and situations in an open, inter-subjective manner' (p. 24). The results of such activity are 'a meditation on life'; geographers would provide more comprehensive mirrors on life experience than is possible for their colleagues from more specialized disciplines, thereby clarifying the structural dynamics of life. Prediction would be impossible, apart for the 'most routinized aspects of experience' (p. 29), but the deeper understanding achieved would allow much more valid social action and planning than is currently possible.

In a later paper, Buttimer (1976) directs geographers' attention to the concept of the lifeworld, that amalgam of the worlds of facts and affairs with those of values which comprise personal experience—'the pre-reflective, taken-for-granted dimensions of experience, the unquestioned meanings, and routinized determinants of behaviour' (p. 281). Positivism is rejected as a method for analysing the lifeworld because it separates the observer from that which he is studying; as a result, he fails to explain the human experience. Idealism is rejected, too, because it accepts that there is a real world outside the individual's consciousness. Phenomenology, on the other hand, is a path to understanding, on which informed planning can be built:

> It helps elucidate how . . . meanings in past experience can influence and shape the present . . . extremely important as preamble not only to scientific procedure, but also as a door to existential awareness. It could elicit a clearer grasp of value issues surrounding one's normal way of life, and an appreciation of the kinds of education and socialization which might be appropriate for persons whose lives may weave through several milieux (p. 289)

The result is an understanding of man's actions as he understands them, rather than in the terms of abstract, outsider-imposed models and theories.

Berry (1973b), too, has backed this phenomenological orientation, calling for

> a view of the world from the vantage of *process metageography*. By metageography is meant that part of geographic speculation dealing with the principles lying behind perceptions of reality, and transcending them, including such concepts as essence, cause and identity (p. 9).

But not all believe that phenomenology can entirely replace the positivist approach. Walmsley (1974), for example, accepts the merits of the case just presented because so many human decisions are based on 'experiential' rather than 'factual' concepts, but feels that the scale of geographical enquiry, and its long tradition of certain types of empirical work, will require maintaining the positivist orientation. That the perceived world is not necessarily the same as the real world will need to be realized, but 'logical consistency and empirical truth will remain central to geographical enquiry provided the importance of values is recognized' (p. 106).

Gregory (1978a), however, is critical of both positivism and phenomenology. Those favouring the former are criticized for 'they remain committed to a positivist epistemology which makes social science an activity performed *on* rather than *in* society, one which portrays society but which is at the same time estranged from it' (p. 51) and for supporting a procedure which, because it so often assumes *ceteris paribus* in testing its models, can never be sure why these fail, when they do, to replicate reality (p. 66). The necessity for hermeneutic approaches is recognized, but these will not be sufficient to provide a satisfactory foundation of themselves, because they ignore the 'constraints on social action which are so much part of the taken-for-granted life-world of the actors' (Gregory, 1978b, p. 166). Thus:

> A geography of the life-world must therefore determine the connections between social typifications of meaning and space-time rhythms of action and uncover the structures of intentionality which lie beneath them (Gregory, 1978a, p. 139).

But

> A major deficiency . . . is [the] restricted conception of social structure: in particular, it ignores the material imperatives and consequences of social actions and the external constraints which are imposed on and flow from them (*ibid*).

Hermeneutics, then, must be incorporated with investigations of those imperatives and constraints; such incorporation produces a critical science, whose nature is discussed in the last chapter.

Whereas a key feature of the positivist/spatial-science approach has been a great numerical superiority of practitioners over preachers, the phenomenological movement (like the idealist) has been characterized by the converse—much preaching and little practice:

> There is an essential difference between the contemplative intentions of this transcendental philosophy and the practical concerns of a social science, so that it is scarcely surprising that . . . geographers' . . . efforts have been directed towards the destruction of positivism as a *philosophy* rather than the construction of a phenomenologically sound *geography* (Gregory, 1978a, pp. 125–6).

There are some examples of the method's use, however. Tuan, in particular, has written several interpretative essays, and Relph, who 'would much prefer to see substantive applications rather than discussions of the possible uses of phenomenology' (1977, p. 178), has published his thesis on *Place and Placelessness* in which—implicitly, he says—phenomenological methods are used 'to elucidate the diversity and intensity of our experiences of place' (Relph, 1976, p. i): his essential themes are the sense of place and identity in the human make-up and the destruction of this through the growing placelessness of modern design. Other work generally quoted as phenomenological includes pieces on European settlement of the New World. Powell (1972), for example, has written on images of Australia and in his major work on the settlement of Victoria's western plains (Powell, 1970) has examined the conflict between official and popular environmental appraisals, the dialogue between these, and the learning process which resulted in the final settlement pattern (see also Powell, 1977). Billinge (1977), however, has queried whether all such works are really phenomenological:

> the idea has spread that since certain branches of our discipline are less susceptible to quantitative reduction (and, so the argument continues, by false extension, to *scientific* analysis), we can justify our partially formulated hypotheses, exploit the atypicality of our data, cease worrying about the validity of our reconstruction and within some weakly articulated framework label the whole exercise phenomenological (p. 64).

The study of perceived environments represents an 'important and vigorous movement' (p. 65) but, Billinge argues, phenomenology is not just the study of such environments: it is an approach to the study of those subjective sources which in its method is presuppositionless, concerned with the human consciousness and not only its output. Thus 'phenomenological we have by no means become' (p. 67). In fact, much of the work which is claimed as phenomenological is probably closer to the idealist position, in that it does not attempt to investigate essences.

Closely associated with phenomenology is existentialism, and geographers arguing for one or the other have encountered difficulties in their separation (Entrikin, 1976): the main difference is that whereas phenomenology is concerned with the foundations of knowledge—the transcendental essences—existentialism focuses on the nature of human existence, on how man knows the world through 'his physical presence, feelings or emotions' (p. 621). Often the two are bracketed as humanistic approaches, however: 'Humanist geographers can best be understood as geographers seeking to regain . . . prescientific awareness of our environment' (p. 625).

Some humanistic geographers set their work in opposition to that of positivists, whereas others only claim a complementary role. Thus Ley (1978) characterizes positivist geographical studies as the 'lonely reflections' of the researcher rather than those of 'men in context', and Ley and Samuels (1978) talk of 'pallid economic man' and his intellectual *rigor mortis*, of the 'mythical glorification of technique', and 'the positivists were able to overawe a generation of geographers whose philosophical reading had rarely passed beyond Hartshorne's *Nature of Geography*' (p. 11). They argue for understanding rather than prediction, for a reconciliation of social science with man. Entrikin (1976), on the other hand, states that:

> humanist geography does not offer a viable alternative to, nor a presuppositionless basis for, scientific geography as it is claimed by some of its proponents. Rather the humanist approach is best understood as a form of

criticism. As criticism the humanist approach helps to counter the overly objective and abstractive tendencies of some scientific geographers (p. 616).

Its criticisms counter the 'dogmatic, abstracted, and narrow' scientific approach by providing descriptions—'of objects as they present themselves to consciousness' (p. 631) and of the non-material aspects of human experience (Glacken, 1978)—rather than by attempting explanations, and by focusing on intuition rather than on data which are sensed. In this way, by investigating attitudes, impressions and subjective relations to places, the human condition is emphasized. Examples of such work include Samuels's (1978) discussion of the creation of existential space, Western's (1978) outline of inter-racial relations in Cape Town, and Gibson's (1978) analysis of social differences between the cities of Vancouver and South Vancouver. But in general, by the late 1970s, one could only conclude, with Hay (1979), that phenomenology and related approaches had been proposed more as a reaction to the shortcomings of positivist studies than as a set of alternative methodologies: whereas the positivist/quantifiers initiated their revolution by inviting emulation, therefore, the humanists have preferred to invite only innovation.

Conclusions

Two very different types of work have been reviewed in this chapter. The common thread linking them is that both are concerned with positive rather than normative investigation, with attempts to uncover how humans behave in the world rather than with contrasts between actual patterns of behaviour in space and those predicted from normative theories. Both types are part of a more general trend towards an anthropocentric focus within the social sciences, which in turn reflects reorientations in the external environment. From the mid-1960s on there was a growing disillusion with science and technology, especially among students, and the popularity of the social sciences boomed. Within the latter, there was a shift in emphasis from the aggregate to the individual, an increase in the relative volume of research conducted at the microscale, and growing unease about the role of social scientists in planning mechanisms. Both behavioural (or behaviourist) geography and the various hermeneutic approaches reflect these trends.

Beyond the general concern with the individual as a decision-maker there is little else to connect the two types of work discussed here. The former, as already pointed out, has maintained strong ties with the positivist/spatial-science tradition; data are collected from individuals, but these almost all concern the conscious elements in action; they are usually aggregated in order to allow statistically substantive and significant generalization to be made about spatial behaviour, almost certainly in the context of the normative models of the spatial science school. In the hermeneutic strand, on the other hand, the intent has been to understand the individual *qua* individual, but the output of substantive studies has not been great. This may indicate that the ethos of the positivist tradition has maintained a major constraint to the wholesale adoption of new philosophies and methodologies. Many investigations in human geography from 1970 onwards involved some behavioural concepts other than those of economic man, but relatively few human geographers have become humanist geographers.

6

Relevance, liberals, and radicals

The late 1960s and early 1970s were traumatic years in the countries being studied here. The underlying problem was economic uncertainty: after two decades of relatively high rates of economic growth and prosperity, the American and British economies began to experience serious difficulties. Further, it became increasingly clear that the prosperity of the previous decades had not been shared by all, and this was highlighted in the United States by the growing tempo of the Civil Rights movement. Student protest erupted in several countries during 1968, much though not all of it related to the Vietnam War, and increasing concern was being expressed about man's despoliation of his physical environment. Other issues included the relatively repressed state of women in western societies (this reached into the geographical profession: Zelinsky, 1973a).

Much of the protest of the late 1960s was centred on particular issues and was relatively ephemeral. For many of those involved, the aim was to win reforms within society, in the classical liberal manner, while leaving its major structure untouched; this was reflected, for example, in the 1972 McGovern campaign for the American presidency. For some, however, disillusion led to what Peet (1977) has termed a 'breaking-off' from liberalism and a move to more radical political stances. As he expressed it:

> The starting point was the liberal political social-scientific paradigm, based on the belief that societal problems can be solved, or at least significantly ameliorated, within the context of a modified capitalism. A corollary of this belief is the advocacy of pragmatism—better to be involved in partial solutions than in futile efforts at revolution. Radicalization in the political arena involved, as its first step, rejecting the point of view that one more policy change, one more 'new face', would make any difference (p. 242).

Eventually, he says, the radicals exploded 'through the thick layers of ideology which in the most dangerous, mass-suicidal way, protect late capitalism' (p. 243) and settled on forms of socialism. Their search for a perspective, and their debates with their more 'liberal' colleagues, form the material for this chapter.

Disenchantment and disillusion in academic geography

A forthcoming revolution in human geography, against the innate conservatism of behavioural studies, was foreseen by Kasperson (1971):

> The shift in the objects of study in geography from supermarkets and highways to poverty and racism has already begun, and we can expect it to continue, for the goals of geography are changing. The new men see the objective of geography as the same as that for medicine—to postpone death and reduce suffering (p. 13).

This shift in objects and objectiveness is not easily defined, however; it is best illustrated by the writings of one senior geographer, Wilbur Zelinsky (born in 1921), who was president of the Association of American Geographers (AAG) in 1973. His views may not be entirely typical, and are certainly more firmly stated than those of his peers, but they reflect a growing disillusion within geographical circles with past achievements (see also Robson and Cooke, 1976) and a wondering about future directions.

Zelinsky's (1970) first statement began:

> This is a tract . . . The reader is asked to consider what I have come to regard as the most timely and momentous item on the agenda of the human geographer: the study of the implications of a continuing growth in human numbers in the advanced countries, acceleration in their production and consumption of commodities, the misapplication of old and new technologies, and of the feasible responses to the resultant difficulties (p. 499).

He developed three basic arguments:
1 that man is inducing for himself a state of acute frustration and a crisis of survival;
2 that these conditions originate, and can only be solved, in the 'advanced' nations; and
3 that the current 'growth syndrome' has profound geographical implications.
Material accumulation can no longer be considered progress, he argues, for it is not sustainable; effort is currently misallocated on a massive scale, and there is a major geographical task involved in its sensible reallocation.

Zelinsky suggests five attitudes towards the problems of the growth syndrome:
1 ignore it;
2 accept that it will occur, eventually;
3 admit the existence of problems, but argue that they are easily solved by the free market, perhaps with state guidance;
4 claim that nothing can be done, but that in any case we will survive; and
5 realize the potential for immediate, unprecedented trouble.

Zelinsky's own attitude is clearly the last of these, and he proposes three roles for the geographer in facing the oncoming disasters. The first—which involves a minimal political commitment and 'should not offend even the most rock-ribbed conservative scholar' (p. 518)—is the geographer as diagnostician, applying 'the geographic stethoscope to a stressful demography' (p. 519), mapping what he calls geodemographic load, environmental contamination, crowding and stress. The second involves the geographer as prophet, projecting and forecasting likely futures. Finally, there is the geographer as architect of utopia, educating with regard to problems and possible solutions and providing support for the unknown leaders who have the political will to guide society through the coming 'Great Transition'.

Ending his 'declaration of conscience' on a pessimistic note, Zelinsky argues regarding geographers,

> how woefully deficient we are in terms of practitioners, in terms of both quantity and quality, how we are still lacking in relevant techniques, but most of all that we are totally at sea in terms of ideology, theory and proper institutional arrangements (p. 529).

His criticisms are not confined to geographers, however; in his presidential address to the Association of American Geographers (Zelinsky, 1975) he applied them to scientists en masse. Science, to Zelinsky, is the twentieth-century religion, but he claims it has failed to avert the oncoming crisis. Its disciplinary specialisms and separatism 'fog perception of larger social realities' (p. 128), while 'fresher, keener insights, along with much better prose' (p. 129) are produced by the brighter contemporary journalists.

Zelinsky identifies five crucial axioms as foundations of science:
1 that the principle of causality is valid for studying all phenomena;
2 that all questions are soluble;
3 that there is a final state of perfect knowledge;
4 that findings have universal validity; and
5 that total scientific objectivity is possible.
But the social sciences have failed to live up to these, he claims, for several reasons: their immaturity; their use as a refuge for mediocre personnel; the difficulty of their subject matter concerning interpersonal relationships; their problems of observation and experimentation; and the political and other problems involved in applying their proposed solutions. But furthermore, their major cause for failure is the irrelevance of the natural-science model to the study of society:

> If we are in pursuit of nothing more than information or knowledge, then there is some value in copying the standard formula of a research paper in the so-called hard sciences. . . . But if we are in pursuit of something more difficult and precious than just knowledge, namely understanding, then this simple didactic pattern has very limited value (p. 141).

Zelinsky's views were echoed in another presidential address to the same body, with Ginsburg (1973) writing that:

> Much so-called theory in geography . . . is so abstracted from reality that we hardly recognize reality when we see it. . . . The increasing demand for rigour to cast light on trivia has come to plague all of the social sciences . . . the most important questions tend not to be asked because they are the most difficult to answer (p. 2).

At about the same time, the Association established a Standing Committee on Society and Public Policy (Ginsburg, 1972), which White (1973) hoped would 'be alert to distinguishing the fatuous problems and the activities that are pedestrian fire-fighting or flabby reform' (p. 103). This statement was made at a session on geography and public policy at the Association's 1970 conference; a similar theme was chosen for the entire 1974 conference of the Institute of British Geographers (IBG). It should be noted, however, that not all geographers accepted White's method— using the Association of American Geographers to influence public policy—whilst not denying the value of geographic method in social engineering. According to Trewartha (1973):

> I must demur when he proposes that it should be a corporate responsibility of our professional society to become an instrument for social change. . . . From the beginning, the unique purpose of the Association of American Geographers has been to advance the cause of geography and geographers; it was never intended to be a social-action organization. . . . All kinds of research, pure as well as applied, should be equally approved and supported by the AAG (p. 79).

Two major arguments can be identified in the materials just reviewed: that geographical research should be relevant to major societal problems; and that the positivist-based spatial-science methodology is inappropriate for such a task. As several have been quick to point out, neither of these concerns was particularly new, especially the first: House (1973), for example, has reviewed the tradition of involvement in public policy by British geographers, and Stoddart (1975b) has identified the late nineteenth-century views of Réclus and Kropotkin as 'the origins of a socially relevant geography' (p. 190)—the latter was later rediscovered by the 'radicals' (Peet, 1978). Nor was the more 'revolutionary' approach of those who had 'broken-off' from liberalism particularly novel. Santos (1974), for example, has reminded English-speaking geographers of the marxist-inspired works of Jean Dresch on capital flows in Africa and of Jean Tricart on class conflict and human ecology, both prior to the Second World War; an English geographer, Keith Buchanan, was well known for his left-wing opinions in New Zealand during the 1960s, and his presidential address on the need for studying 'the absolute geographical primacy of the state; especially in the non-Western world' (Buchanan, 1962), produced an acid response from Spate (1963).

Relevance, on what, and for whom?

The claims that geographical work should be more relevant to major societal problems raised queries about the nature of that relevance, and it soon became apparent that there was no consensus on what should be done, and why. The ensuing debate is illustrated by a number of contributions to the British journal *Area* during the early 1970s.

The opening statement was by Chisholm (1971b) who identified differences between governments, with their interests in cost-effective research and their primacy in decision-making, and academics, some of whom are concerned to protect their academic freedom and their right to be the sole judges of what they study and publish. Traditionally, geographers had advised governments as information gatherers and 'masterful synthesizers' and had not been involved in the final stages of policy-making: they had been delvers and dovetailers, but not deciders. On the latter role, unfortunately:

> The magic of quantification is apt to seem rather less exciting when the specifications of the goods it can deliver are inspected at close quarters (p. 66).
> . . . The danger with empirical science is the absence of guidance at the normative level as to which of various options one should take (pp. 67–8).

The challenge to human geography, according to Chisholm, was to define such norms.

To Eyles (1971), the focus of relevant research should be 'some of the social and spatial inequities in society' (p. 242), and the challenge is to study the distribution of power in society, which is the mechanism that allocates scarce resources. Research would then identify the disadvantages of relative powerlessness—and would provide the basis for policy which redistributes resources.

Also in 1971, British readers were introduced to the ongoing debates in American geography with two reports on the 1971 Association of American Geographers' conference. This was not the first at which major social issues had been raised: the 1969 meeting should have been in Chicago, but it was transferred to Ann Arbor as a protest over events at the 1968 Democrat Party Convention in Chicago. At Ann Arbor, the radical journal *Antipode* was launched, and those attending were confronted by the problems of the inhabitants of Detroit's black ghetto. By the 1971 meeting, 'many geographers were deeply frustrated by a sense of failure' (Prince, 1971b, p. 152) to deal with major social issues, but while some members were taking notice of 'the sufferings of the outside world' (p. 152) other scholars remained 'locked in private debates, preoccupied with trivia, mending and qualifying accepted ideas' (p. 153).

In his report of the 1971 conference, Smith (1971) suggested that American geography was about to undergo another revolution, to counter a situation in which 'geography is overpreoccupied with the study of the production of goods and the exploitation of natural resources, while ignoring important conditions of human welfare and social justice' (p. 154). This forthcoming revolution would involve fundamental revaluation of research, teaching activities, and basic social philosophies, and was represented at the conference by activities such as the foundation of SERGE (The Socially and Ecologically Responsible Geographer) by Zelinsky and others and by a motion at the Annual General Meeting condemning United States involvement in Vietnam. Smith was unsure, however, whether this revolution, with its emphasis on social as against economic concerns, would spread to Britain:

> The conditions which have helped to spawn radical geography in the United States include the existence of large oppressed racial minorities, inequalities between rich and poor with respect to social justice, a power structure and value system largely unresponsive to the needs of the underprivileged, and an unpopular war which is sapping national economic and moral strength. These conditions do not exist in Britain or exist in a less severe form, and the stimulus for social activism in geography is thus considerably less than in America (pp. 156–7).

Dickenson and Clarke (1972) responded that British geographers had long been concerned with 'relevant' issues, with particular respect to the Third World.

Another commentary on the 1971 conference of the AAG at Boston came from Berry (1972c) who felt that he had observed just a new fad involving 'new entrants to the field seeing their "turf" '. He could identify no real commitment:

> The majority of the new revolutionaries, it seems, are essentially 'white liberals', quick to lament the supposed ills of society and to wear their bleeding hearts like emblems or old school ties—and quicker to avoid the hard work that diagnosis and action demand. A smaller group of hard-line marxists keeps bubbling the potage of liberal laments. *In neither group is there any profound commitment to producing constructive change by democratic means. . . . If either of these will be the 'new geography' of the 1970s, count me out* (pp. 77–8).

To him, the academic geographer should provide a base of knowledge on which policy can be built, and this involves a close involvement with the education of future generations of policy-makers. But to Blowers (1972), 'The issue is not how we can cooperate with policy-makers, but whether and in what sense we should do so. It is a question of values' (p. 291). To him, the activities proposed by Berry would be strongly supportive of the status quo, and unlikely to produce fundamental social reform. Smith (1973a) responded to Berry that 'bleeding hearts sometimes help to draw attention to important issues, and marxists can make valuable contributions in the search for alternatives to existing institutions and policies' (p. 1) and pointed out that the current 'fad' was no more pronounced than that of the quantifiers a decade earlier. The results of that earlier 'revolution' offered little for the solution of social problems, however, and Smith doubted the value of the large projects established by the AAG as part of its geography and public-policy drive. Instead the research focus should highlight particular problems, while teaching should have emphases on 'a man in harmony with nature rather than master of it, on social health rather than economic health, on equity rather than efficiency, and on the quality of life rather than the quantity of goods' (p. 3). Chisholm (1973) advocated caution in the corridors of power, because geographers had done insufficient substantive research to back up a 'hard sell'; Eyles (1973, p. 155) argued that any entry to those corridors 'assumes that the structure underlying policy alternatives is basically sound'; and Blowers (1974) wrote that in the corridors one can only influence, not decide, and that for the latter task geographers must develop their political convictions and act accordingly.

This debate on if, and how, geographers should contribute to the solution of societal problems was a major item at the 1974 annual conference of the IBG. In his presidential address, Coppock (1974) presented the challenges, opportunities and implications of geographical involvement in public policy, an involvement which he felt the current generation of students welcomed. Policy-makers, he felt, were largely ignorant of the potential geographical contribution, while at the same time geographers seemed unaware that 'there is virtually no aspect of contemporary geography which is not affected to some degree by public policy' (p. 5). Coppock sought to change this, to have geographers identify the contributions that they could make, to encourage research relevant to those contributions, and to enter a dialogue with those who advise on and implement public policy.

Other conference contributors were neither as optimistic nor as committed as Coppock. Hare (1974), himself an adviser to the Canadian government, reacted to cries that geographers were not consulted enough with the reply 'Thank goodness'. His argument was that geography, as a discipline, is irrelevant to the separate domain of public-policy-making, although geographers, as individuals, because of the breadth of their training, could offer much that was valuable: his conclusion was 'Geography no, geographers yes' (p. 26). His was a different response to mounting social concerns to that of Steel (1974) who told the British Geographical Association:

> As geographers we often get hot under the collar over the number of theoretical economists who are called on to advise the governments of developing countries. We comment on how much better World Bank surveys of countries would be if they were prepared, at least in part, by geographers. . . . We wonder why university departments of geography are not engaged on a consultancy basis more often than they are, and we marvel that the Overseas Development Administration in London has only a handful of geographers on its staff where, we feel, an army would be more appropriate (p. 200).

In a later paper, Hare (1977) argues that a major reason for a lack of geographical contributions to public policy in recent years may be the poverty of their training: in recent years we have 'swept geography departments into the social-science divisions of faculties of arts and sciences where, from playing second fiddle to geologists or literary critics, we learned to play second fiddle to economists and sociologists' (p. 263). Geographers, he argued, would have to rebuild their discipline based on the centrality of man-environment interactions, with a new brand of physical geography that leans heavily on biological ideas and sources. Thus:

> We must reassert the old, essential truth that geography is the study of the earth as the habitat of man, and not some small sub-set of that gigantic theme (p. 266)

and regional syntheses should be stressed, because

> Realizing that regional perspectives are necessary in politics is to get one's manhood back (p. 269).

Hare's views on the current irrelevance of geography to public policy were supported by Hall (1974) who argued that

> geography, most clearly of all the social sciences, has neither an explicit nor an implicit normative base . . . spatial efficiency . . . is rather a description of what men seek to do in actuality . . . not . . . any objective to be achieved or objective function to be maximized. (p. 49).

Policy-makers must seek their norms elsewhere; geographers, meanwhile, must develop a new political geography which will aid in their understanding of the crucial role of political decisions in structuring spatial systems (Johnston, 1978b).

Two other papers given at the 1974 IBG conference argued against Coppock's programme. Leach (1974), for example, claimed that governments, as paymasters, already constrained what geographers could do research on, and as a result geographers were being used; their only alternative was political action. Harvey's (1974c) contribution was entitled 'What kind of geography for what kind of public policy?'. Individuals wishing to become involved in policy-making were, he argued, stimulated by motives such as personal ambition, disciplinary imperialism, social necessity, and moral obligation: at the level of the whole discipline, on the other hand, geography had been coopted, through the universities, by the growing corporate state, and geographers had been given some illusion of power within a decision-making process designed to maintain the status quo. Indeed, to Harvey the corporate state is 'proto-fascist' (p. 23), a transitional step on the path to the barbarism of Orwell's *1984*. The function of academics, he claimed, was to counter such trends, to expunge the racism, ethnocentrism and condescending paternalism from within their own discipline and to build a humanistic subject and thereby assist all human beings 'to control and enhance the conditions of our own existence' (p. 24).

Harvey (1973) had argued previously that the current mode of analysis in geography offered little for the solution of pressing societal concerns:

> There is an ecological problem, an urban problem, an international trade problem, and yet we seem incapable of saying anything of depth or profundity about any of them. When we do say anything, it appears trite and rather ludicrous. . . . It is the emerging objective social conditions and our patent inability to cope with them which essentially explains the necessity for a revolution in geographic thought (p. 129).

He recognizes three types of theory:

1   status quo, which represents reality accurately but only in terms of static patterns, and therefore cannot make predictions which will lead to fundamental social change;

2   counter-revolutionary, which also represents reality, but obfuscates the real issues—it is 'a perfect device for non-decision making, for it diverts attentions from fundamental issues to superficial or non-existent issues' (p. 151)—as, he would undoubtedly claim, with much of the quantitative work of the 1960s; and

3   revolutionary, which is grounded in the reality it seeks to represent, and is formulated so as to encompass the contradictions and conflicts which produce social change.

Harvey clearly wished to write revolutionary theory, thereby over-throwing the current paradigm; his blueprint for geography

> does not entail yet another empirical investigation. . . . In fact, mapping even more evidence of man's patent inhumanity to man is counter-revolutionary in the sense that it allows the bleeding-heart liberal in us to pretend we are contributing to a solution when in fact we are not. This kind of empiricism is irrelevant. There is already enough information. . . . Our task does not lie here. Nor does it lie in what can only be termed 'moral masturbation' of the sort which accompanies the masochistic assemblage of some huge dossier of the daily injustices. . . . This, too, is counter-revolutionary for it merely serves to expiate guilt without our ever being forced to face the fundamental issues, let alone do anything about them. Nor is it a solution to indulge in that emotional tourism which attracts us to live and work with the poor 'for a while'. . . . These . . . paths . . . merely serve to divert us from the essential task at hand.
>
>   This immediate task is nothing more nor less than the self-conscious and aware construction of a new paradigm for social geographic thought through a deep and profound critique of our existing analytical constructs. This is what we are best equipped to do. We are academics, after all, working with the tools of the academic trade . . . our task is to mobilize our power of thought, which we can apply to the task of bringing about a humanizing social change (pp. 144-5).

To Harvey, then, relevant geography involves the revision of geographic theory; such new theory will be built on a marxist base, and its dissemination will achieve social reform through the education process. (This gradualist view to reform through education is only implicit in Harvey's work; Johnston, 1974, p. 189.)

This section has indicated a growing polarization of views within the geographical community during the early 1970s, largely between those advocating a liberal approach to social problems with geographers providing information and procedures for their amelioration or solution within the existing, capitalist framework, and those arguing for more radical reappraisals of the discipline's role, based on socialist philosophies. (There were also, of course, those who were unconvinced of the need to leave the academic ivory tower.) The major contributions from these two groups are reviewed in the next sections.

The liberal contribution

In the present context, liberalism is defined as the idea 'shared by many modern American liberals who, characteristically, combine a belief in democratic capitalism with a strong commitment to executive and legislative action in order to alleviate social ills' (Bullock, 1977, p. 347). Liberals, then, are concerned that all members of a society do not fall below certain minimum levels of well-being (variously defined), and are prepared for state action within the capitalist structure in order that this can be achieved. Within geography, much of the work conducted in this ethos has focused on description rather than on theory-construction: the few exceptions include Chisholm's (1971a) investigation of the potential of welfare economics as a basis for normative theory which does not involve profit-maximization goals (see also Wilson, 1976b).

Mapping welfare
During the 1960s, a lot of research was reported, under the general title of factorial ecologies (p. 73), on the application of multivariate statistical procedures to large data matrices representing spatial variations in population characteristics. These works, according to Smith (1973b, p. 43), were over-reliant on certain types of census data and therefore provided little information on social conditions. There had been some earlier attempts to structure analyses of such data towards particular ends, as in the work of rural sociologists on farmers' levels of living (Hagood, 1943), a concept introduced to the geographical literature by Lewis (1968), and Thompson *et al*.'s (1962) investigation of variations in levels of economic health between different parts of New York state. Only in the 1970s, however, was the factorial ecology set of procedures adapted to the task of mapping social welfare on any significant scale.

Two workers led in this adaptation. For Knox (1975), a fundamental objective for geography is to map social and spatial variations in the quality of life, both as an input to planning procedures and as a means of monitoring policies aimed at improving welfare. The concept of level-of-living was divided into three sets of variables for this reporting task—physical needs (nutrition, shelter, and health); cultural needs (education, leisure and recreation, and security); and higher needs (to be purchased with surplus income)—and statistical procedures were used to provide accurate portrayals of the spatial variations in meeting these needs. With the resultant maps, the geographer must then decide whether he is playing a sufficient role in awakening human awareness of the extent of the disparities or 'whether he is under an obligation to help society improve the situation' (p. 53).

Smith's (1973b) very similar work was set in the context of the American social-indicators movement and the growing belief there that GNP and national income 'are not necessarily direct measures of the quality of life in its broadest sense' (p. 1). His aim was to initiate the collection and dissemination of territorial social indicators, to point out the extent of discrimination by place of residence which occurs in the United States. Again, multivariate statistical procedures were employed to provide the needed maps, at inter-state, inter-city and intra-urban scales. With these, it was necessary for geographers to get involved in the political struggle, to be social activists as well as social reporters.

To use Chisholm's terms (p. 146), these two works represented the geographer as delver and dovetailer, as a provider of information on which more equitable social planning could be based. Several other studies of individual elements in that planning performed similar roles, and also suggested spatial policies which could lead to social improvements. Harries (1974), for example, studied spatial variations in crime rates and the administration of justice, and argued that predictive models of criminal patterns, in the positivist mould, could aid in the organization of police activities; Shannon and Dever (1974) investigated variations in the provision of health-care facilities and argued for spatial planning which would improve the services offered to the sick (which is different from a geography of prophylaxis: Fuller, 1971); and Morrill and Wohlenberg (1971) studied the geography of poverty in the United States, proposing both social policies—higher minimum wages, guaranteed incomes, guaranteed jobs, and stronger anti-discrimination laws—and spatial policies—an extensive programme of economic decentralization to a network of regional growth centres—which would alleviate this major social problem.

An alternative, and highly personal, programme of mapping variations in human welfare was advanced by Bunge, who had prepared a 'geobiography' of his home area, part of the black ghetto of Detroit (Bunge, 1971). His is a deeply humanitarian concern for the future of mankind, which he interprets as a need to ensure a healthy existence for children; he wants a 'dictatorship of the children' (Bunge, 1973b, p. 329) with regions—'may the world be full of happy regions' (p. 331)—

designed for them. This requires a reduction in the worship of machines, which are inimical to children's health (Bunge, 1973c) and a mapping of the sorts of variables never collected by external agencies and therefore requiring the development of geographical expeditions within the world's large cities; these maps would include roach regions, parkless spaces, toyless regions, and rat-bitten-children regions (Bunge, 1973d), and some have been prepared for Detroit and for Toronto (Bunge and Bordessa, 1975).

Attempts at understanding
The mapping investigations just discussed are very largely descriptive, and any prescriptions offered are based on limited theoretical foundations. Other investigations included attempts to develop the necessary theoretical understanding. Cox (1973), for example, looked at the urban crisis in the United States—the racial tensions and riots, municipal bankruptcies, and the role of the government in the urban economy— presenting his analysis in terms of conflict over access to sources of power. This was intended as part of an educational exercise, for:

> It would be utopian to think that we can propose solutions on the basis of our analysis. The locational problems and locational consequences of policies weave too intricate a web for that to be possible. All we can hope to do is inform. To be aware of the problems and of their complexity may induce some sensitivity in a citizenry which has shown as yet precious little tolerance for the other point of view (p. xii).

Nevertheless, his final chapter is entitled 'Policy implications' in which are discussed two imperatives towards greater equity in the provision of public services—the moral imperative and the efficiency imperative (the latter applies to the total level of welfare in society as well). The policies presented involve spatial reorganization to achieve the desired equity, including metropolitan integration, community control, population redistribution, and transport improvements. A similar focus on spatial reorganization is provided in Massam's (1975) review of geographical contributions to social administration: his evaluation of service provision is in terms of the spatial variables of distance and accessibility, with major chapters on the size and shape of administrative districts and on the efficient allocation of facilities within such districts (see also Hodgart, 1978).

Cox's work heralded an increased geographical interest in a much neglected field, the role of the state in capitalist society (see also Cox, Reynolds and Rokkan, 1974; Dear and Clark, 1978). Traditionally, political geography had been concerned with the state at the macro-scale only, dealing with political regions and boundaries and with the operations of the international political system (e.g. Muir, 1975). The similarly underdeveloped field of electoral geography had highlighted spatial variations in voting, but there had been little work on both the geographical inputs to voting and the geographical consequences of the translation of votes into political power (Taylor, 1978; Taylor and Johnston, 1979). The state is involved in many aspects of economic and social geography, however, as both Buchanan (1962) and Coppock (1974) have stressed, but few geographers have investigated this involvement in any detail, or the electoral base on which it is founded (Brunn, 1974; Johnston, 1978b).

A more wide-ranging attempt to present an understanding of spatial variations in well-being was presented by Coates, Johnston and Knox (1977). After defining the components of well-being and mapping their variations at three scales—international, intra-national, and intra-urban—they introduced three sets of causes of such variations: the division of labour; accessibility to goods and facilities; and the political manipulation of territories. Finally, they evaluated various spatial policies aimed at the reduction of spatial inequalities, such as various forms of positive discrimination by areas. Their conclusions were that, of their three sets of causes, the division of labour is the primary determinant of levels of social well-being. Creation of this division is a social and not a spatial process, though it has clear spatial consequences, so that:

> The root causes of spatial inequalities cannot be tackled by spatial policies alone, therefore. Inequalities are products of social and economic structures, of which capitalism in its many guises is the predominant example. Certainly inequalities can be alleviated by spatial policies . . . but alleviation is not cure: whilst capitalism reigns, however, remedial social action may be the best that is possible . . . the solution of inequalities must be sought in the restructuring of societies (pp. 256–7).

Hägerstrand (1977) was drawn to similar, if not more pessimistic, conclusions:

> When the world is stable and/or unhampered liberalism prevails, then there is probably not much to do for geographers except surviving in academic departments trying to keep up competence and train schoolteachers in how wisely arranged the world is (p. 329).

(Hägerstrand's definition of liberalism in this case is undoubtedly that of economic liberalism, which favours price competition in the 'market-place' and frowns on 'state interference': Brittan, 1977, p. 188).

A more ambitious attempt at explaining spatial variations in well-being is Smith's (1977) book, in which he argues that:

> the well-being of society as a spatially variable condition should be *the* focal point of geographical enquiry . . . if human beings are the object of our curiosity in human geography, then the quality of their lives is of paramount interest (pp. 362–3)

and he acts on this by essaying

> a restructuring of human geography around the theme of *welfare* . . . to

provide both positive knowledge and guidance in the normative realm of evaluation and policy formulation (p. ix).

The book proceeds from theory through measurement to application. The theoretical section is an amalgam of normative welfare economics with marxian perspectives on the creation of value plus the political conflict for power.

> The analysis will inevitably reveal certain fundamental weaknesses of the contemporary capitalist-competitive-materialistic society, but the temptation to offer a more radical critique of existing structures has been resisted, in favour of an approach that builds on the discipline's established intellectual tradition (p. xi).

Two types of solution to perceived spatial inequalities are identified: liberal intervention, and radical, structural reform—the former is emphasized because 'most social change is incremental rather than revolutionary' (p. xii). Thus one concluding chapter—entitled 'Spatial Reorganization and Social Reform'—emphasizes reorganization of administrative areas, although geographers are later urged to be 'open-minded enough to see that there may be greater merit to some of the development strategies practised with evident if not painless success in the USSR, China, Cuba and so on' (p. 358). Further, counter-revolutionary possibilities are foreseen—'over-preoccupation with spatial reorganization may serve the interests of the existing rich and powerful by helping to obscure the more fundamental issues' (p. 359)—and some socialist arguments are forthcoming—'That spaces and natural resources should be privately owned, with their use subject to the chances of individual avarice, altruism or whim, is increasingly an anachronism' (p. 360). But the conclusion is almost phenomenological:

> As geographers we have a special role—a truly creative and revolutionary one—that of helping to reveal the *spatial* malfunctionings and injustices, and contributing to the design of a spatial form of society in which people can be really free to fulfil themselves. This, surely, would be progress in geography (p. 373).

Environmentalism

The late 1960s was a period of rapidly increasing concern about environmental problems; as Mikesell (1974) described it for the United States:

> Towards the end of the 1960s the American public was overwhelmed with declarations of an impending environmental crisis. . . . Since that time, crisis rhetoric and a yearning for simple answers to complicated questions has given way to a more sophisticated and deliberate search for environmental understanding. Ecology has been institutionalized (p. 1).

Two of the leaders of the public debate disagreed as to the cause of the problems (O'Riordan, 1976, pp. 65–80): Ehrlich argued for the primacy of population growth, and popularized the concept of zero population growth; Commoner claimed that technological advances and the consequent rapid depletion of resources plus deposition of pollutants created the major problems.

Both of these arguments have clear geographical components, as Zelinsky and others realized (p. 143), and geographers have a considerable record of activity in resource conservation. In the United States, for example, George Perkins Marsh had written on the topic in 1864 (Lowenthal, 1965) and in the 1930s the climatologist Warren Thornthwaite had been closely involved in the soil-conservation movement established as a consequence of the Dust Bowl phenomenon. Their interest in landscape modification was advanced by Sauer and his followers, and reflected in the symposium *Man's Role in Changing the Face of the Earth*: there was similar interest elsewhere, as exemplified by Cumberland's (1947) pioneering classic on soil erosion in New Zealand. Nevertheless, Mikesell wrote that 'developments in geography have been such that the several phases of national preoccupation with environmental problems have not produced a general awareness of our interests and skills' (p. 2).

As part of the Association of American Geographers' increased commitment to public affairs, its Commission on College Geography established a Panel on Environmental Education and sponsored a Task Force on Environmental Quality. The latter reported (Lowenthal *et al.*, 1973) that geographers would make excellent leaders for the educational tasks in hand because:

1 of the breadth of their training and their ability to handle and synthesize material from a range of sources;
2 their acceptance of the complexity of causation;
3 the range of information which they are trained to tap;
4 their interest in distributions; and
5 their long tradition of study in this area.

All of these had allowed the development of expertise in work on environmental perception, on vegetation succession, and on relationships between land use and soil erosion, which could be used as the bases for environmental impact statements, the elaboration of environmental choices, and international research collaboration.

Geographical efforts in the area of man-environment interrelationships were of two types. The first, traditionally geographical, was concerned with description and analysis. Review volumes such as *Perspectives on Environment* (Manners and Mikesell, 1974) were prepared, and a particular interest in problems of the physical environment of urban areas was generated (Detwyler and Marcus, 1972; Berry and Horton, 1974; Berry *et al.*, 1974). The second type of effort focused more precisely on issues of environmental management (O'Riordan, 1971a, 1971b), with particular emphasis on its economic aspects: as Kates

(1972, p. 519) pointed out, economics provided the theories and pre-scriptions of the 1960s. A topic of especial interest was the study of leisure, of the growing demand for recreational facilities, and the impact of recreational activities on the environment (Patmore, 1970).

Despite such activity, Mikesell concluded in 1974 that the geo-graphical contribution to environmentalism had not been great. Regard-ing the prognostications of *The Limits of Growth* (Meadows *et al*., 1972), for example, he commented that 'the debate on this most relevant of all issues has attracted remarkably little attention from geographers' (Mikesell, 1974, p. 19)—though see Eyre (1978)—and he concluded more generally that: 'one must add hastily that many of the environmen-tal problems exposed in recent years and also many of the social and philosophical issues debated during the environmental crusade have not been given adequate attention by geographers' (p. 20). Such a conclusion is supported by perusal of the contents of recent geographical journals, and of O'Riordan's (1976) lengthy bibliography.

The breadth of study in modern environmentalism is indicated by O'Riordan's volume. Much of the work is set in the liberal humanitarian tradition already illustrated in this section. O'Riordan draws four con-clusions: environmentalism challenges many aspects of Western capital-ism; it points out paradoxes rather than clear solutions; it involves a conviction that better modes of existence are possible; and it is a politiciz-ing and reformist movement, based on a realization of the need for action in the face of impending scarcity and a lack of faith in the western democracies (pp. 300–1). A new social, environmental order is required. O'Riordan identifies three possibles—centralized, authoritarian, and anarchist; his choice is the liberal one, for the middle-of-the-road:

> we must individually and collectively seize the opportunities of the present situation to end the era of exploitation and enter a new age of humanitarian concern and cooperative endeavour with a driving desire to re-establish the old values of comfortable frugality and cheerful sharing (p. 310).

This new era, which will involve a new political order based on a combination of local self-determination and supranationalism, can be achieved through education—environmental education will form a prep-aration for citizenship.

Radical alternatives

Peet (1977) has pointed out that the early 'radical' work by geographers in the late 1960s was liberal in its attitudes (p. 142):

> Radicals investigated only the surface aspects of these questions—that is, how social problems were manifested in space. For this, either we found the conventional methodology adequate enough or we proposed only that 'exist-ing methods of research must be modified to some extent if they are to serve the analytic and reconstructive policies of . . . radical applications' (Wisner,

1970, p. 1) . . . we were fitting into an established market . . . we were amenable to established ways of thinking . . . we were useful in providing background ideas for the formulation of 'pragmatic' public policy directions, and so could not, and were not, engaging in radical analysis and practice (p. 245).

This was illustrated by his own paper on poverty in the United States (Peet, 1971) in which, like Morrill and Wohlenberg (see p. 152), he argued for a series of growth centres in the poverty areas, and also by the tenor of the articles in early issues of the radicals' journal, *Antipode*. Morrill (1969), for example, argued

against the 'New Left' premise that a revolution is the only route to progress . . . the dreams of revolution are naive . . . the New Left vastly exaggerates potential support . . . a 'revolutionary program' is hopelessly dated and simplistic . . . the 'New Left' underestimates the capacity of our society for change. . . . All revolutions seem to have been betrayed by incompetents who preferred exercising power to executing reform (pp. 7–8)

and later (Morrill, 1970b) that

A simple marxist-type change in the ownership of business from private to a government (or union) bureaucracy would in all probability decrease production, and would not necessarily bring any improvement in basic conditions. The key is to retain the institution of private property while instituting social control over its exchange and circumscribing its power over people (p. 8).

The case for a marxist approach was first presented formally by Folke (1972) in a critique of Harvey's (1972) paper on ghetto formation and counter-revolutionary theory. To Folke, geography and the other social sciences are 'highly sophisticated, technique-orientated, but largely descriptive disciplines with little relevance for the solution of acute and seemingly chronic societal problems . . . theory has reflected the values and interests of the ruling class' (p. 13). Liberal arguments such as Morrill's are dismissed as unlikely to succeed. Morrill had argued in his two papers in *Antipode* for change to be brought about by persuasion, producing a capitalist-socialist convergence. But this is the social democrat method practised in Sweden

where it has been shown over and over again that the idea of equal influence for employers and employees is an illusion. After half a century of social-democratic rule injustices and inequalities still prevail. . . . No small group of experts can accomplish anything . . . when it runs counter to the interests of the dominant social forces. These are not interested in equality or justice, but in profit (Folke, 1972, p. 15).

Radical change requires mass mobilization, and so, to Folke, Harvey's call for a new paradigm within geography was insufficient. What is needed is a new paradigm for a unified social science, containing geography, which deals with problems in all their complexity and provides not only theory but the basis for action: 'Revolutionary theory without revolutionary practice is not only useless, it is inconceivable . . . practice is the ultimate criterion of truth' (p. 7).

The major contribution to the case for a marxist-inspired, materialist theory-development within geography was made by David Harvey, notably in his book of essays *Social Justice and the City* (Harvey, 1973). The book is presented as autobiographical, illustrating the evolution of Harvey's views towards an acceptance of Marx's analysis:

> as a guide to enquiry . . . I do not turn to it out of some *a priori* sense of its inherent superiority (although I find myself naturally in tune with its general presupposition of and commitment to change), but because I can find no other way of accomplishing what I set out to do or of understanding what has to be understood (p. 17).

The first part of the book is entitled 'Liberal Formulations' and comprises essays which analyse problems of inequality within societies in terms of the mechanisms which allocate income; the role of accessibility and location in those mechanisms is stressed. This leads him to an attempt to define territorial social justice, which separates the processes allocating incomes from those which produce them. Only in the second part of the book—'Socialist Formulations'—is it

> finally recognized that the definition of income (which is what distributive justice is concerned with) is itself defined by production. . . . The collapse of the distinction between production and distribution, between efficiency and social justice, is a part of that general collapse of all dualisms of this sort accomplished through accepting Marx's approach and technique of analysis (p. 15).

The transition in Harvey's approach is marked by the paper on the ghetto (Harvey, 1972). He begins with a critique of Kuhn's model of scientific development (p. 15), asking how anomalies to the current paradigm arise and how they are translated into crises. The problem with Kuhn's analysis, he claims, is that it assumes that science is independent of its enveloping material conditions, when in fact it is very much geared to its containing and constraining society. Recognition of this point is important for geographers because:

> the driving force behind paradigm formulation in the social sciences is the desire to manipulate and control human activity in the interest of man. Immediately the question arises as to who is going to control whom, in whose interest is the controlling going to be exercised, and if control is going to be exercised in the interest of all, who is going to take it upon himself to define that public interest? (Harvey, 1973, p. 125).

To Harvey (1973), Marxist theory provides

> the key to understanding capitalist production from the position of those *not* in control of the means of production . . . an enormous threat to the power structure of the capitalist world (p. 127).

It not only provides the understanding of the origins of the present system, with its many-faceted inequalities, but also propounds alternative practices which would avoid such inequalities:

> we become active participants in the social process. The intellectual task is to identify real choices as they are immanent in an existing situation and to devise ways of validating or invalidating these choices through action (p. 149).

In such a context, geography can no longer be simply academic, isolated in its 'ivory towers'. Its practitioners must become politically aware and active, involved in the creation of a just society which involves replacement of, not reform of, the present one. The remainder of Harvey's (1973) book does not make this commitment clear, however; it contains one essay on land use and land-value theory, investigating the difficult concept of rent, and another on the nature of urbanism, presenting a marxist interpretation of the process of urbanization.

Peet, too, moved from a liberal to a marxist position, replacing his earlier paper on poverty (Peet, 1971) by a marxist interpretation (Peet, 1975a) based on the assumption that inequality is inherent in the capitalist mode of production. This leads him to a 'metatheory dealing with the great forces which shape millions of lives' (p. 567) within which

> Environmental, or geographic, theory deals with the mechanisms which perpetuate inequality from the point of view of the individual. It deals with the complex of forces, both stimuli and frictions, which immediately shape the course of a person's life (pp. 567–8).

Such environmental resources act as constraints because they define the milieux within which individuals are socialized and presented their opportunities for participation within the capitalist system. Mere redistribution of income through liberal mechanisms, based on taxation policies, will not solve the problems of poverty, therefore; according to Peet, the requirement is for geographers to seek alternative environmental designs, with removal of central bureaucracies and their replacement by anarchistic models of community control. (Harvey—1973, p. 93—disagrees with the latter, pointing out that unless resources are equalized among communities and territories, community control will only result in 'the poor controlling their own poverty while the rich grow more affluent from the fruits of their riches'.)

The marxist alternative, then, proposes a complete break from the approaches espoused by human geographers since the Second World War. Logical positivism is denounced because it can only describe. It argues from an analogy with the natural sciences that there are 'objective facts' which can be observed and analysed through the procedures of the scientific method, not realizing that facts are defined by theories, as argued by those proposing hermeneutic alternatives, and that the ruling ideas in a society are those of its ruling classes. Since their theories are grounded in the status quo, descriptions obtained through logical-positivist procedures cannot be used as a basis for change, nor indeed for understanding change, so that for those who wish to see major societal restructurings towards greater equity, logical-positivist theories are counter-revolutionary. These do not represent the conflicts between individuals that produce differences in life chances, nor do they indicate that space is manipulated by the powerful in order to maintain inequalities.

The alternatives to logical positivism discussed in Chapter 5 are similarly dismissed for ignoring the vital role of the elite in society as the manipulators of its whole structure and the creators of its ideology (Anderson, 1973). By reducing the focus of their study to the individual, geographers ignore the constraints placed on individual freedom of action by that structure and its ideology. Thus, according to Wagner (1976) 'Existentialism and its close relatives in the Phenomenological camp seem to abdicate concern with the processes of history and the wider panoramas of geography, and so perhaps lack direct relevance' (p. 84). The degree of constraint on individuals varies somewhat, according to their class position, and Reiser (1973) concluded from his work on environmental attitudes among a working-class population in London's dockland that 'the *objective* constraints were so great that the elucidation of constraints in subjective knowledge of the environment became meaningless' (p. 53). He felt that geographers, along with other social scientists, had adopted a psychologism which

> mystifies reality by complex exposition of simple isolated problems; it obscures the more obvious and economic conditions operating independently of the individual; and it limits understanding of social change through its ahistorical approach and the frequent focus on trivial matters (p. 54).

Concern for the individual's views, and with their use as a basis for social planning, means that 'the hard realities of the world tend to get swamped in wishful thinking' (p. 55), calibrating models of false consciousness.

Structuralism and other radical alternatives
Not all of the radical contributions to the literature of the 1970s have been, overtly at least, marxist. Eyles (1974), for example, presents a more restricted critique than does Harvey. He dismisses the functionalist approach of logical positivism and the reductionism of the behavioural approaches, and instead argues for a greater emphasis on study of the conflict which is endemic to social life. There are two main models of modern western society:

> one based on consensus, the other on conflict. The first would stress that social life is based on cooperation and reciprocity with norms being the basic element. Such societies would also be characterized by cohesiveness, integration and persistence and life within them would depend on the recognition of legitimate authority, mutual commitments, and solidarity. The conflict model sees sectional interests as being the basis of social life which is,

therefore, divisive, involving inducement and coercion as well as generating the structural conflicts . . . [which are] the central element in a social system (p. 39).

Conflict and power are the key elements in the second model, but much social geography emphasizes 'the functional element at the expense of the disruptive' (p. 45), thereby ignoring the social reality of, for example, the competition for housing.

Eyles argues that geographers should focus on understanding how scarce resources are allocated, and should realize that power is the key to such allocations. The pluralist view of power is that all voters and voting groups are equal, but this is a false view of society, as analysis of housing markets (see below) indicates. Society has an elite dominated by business interests, to which those holding political office are likely to defer, and the activities of that elite determine the geography of opportunities for the majority of society's members. Such power is associated with wealth and the maintenance of inequalities; its ethic of competition

> implies winners and losers and the winners in economic competition will be those with the powers of ownership and control in the productive process. In this way, growth under such a system will most probably lead to greater inequalities. It would seem obvious, therefore, that poverty and the distribution of real income in a spatial system cannot be understood without reference to power and inequality (p. 64).

The marxian mode of analysis is one form of a method of analysis which is widely known as structuralism, although, as Gregory (1978a, p. 115) notes, there have been attempts to dissociate marxian methodology from the structuralist umbrella. In brief, the structuralist argument is that observable phenomena are particular outputs (or realizations) of a given set of mechanisms—or deep structures: a prime example is the anthropology of Claude Lévi-Strauss, which proposes the existence of certain universal deep structures, operating at the neural level, to govern, for example, incest taboos, language forms, and attitudes to foods in all societies (Leach, 1974). Gregory (1978a, pp. 99ff) has illustrated this argument in a geographical context using Lévi-Strauss's analogy of the cam-shaft driven by a machine for cutting the outline of pieces in jig-saw puzzles (Figure 6.1). The deep structure is the governing mechanism—here termed the spatial schema—which is reflected, via a series of transformations, in a series of 'imperatives' or collective driving forces—the spatial structure. Finally, the cam-shaft connects these imperatives to the observable world, operating within the constraints of the structure, to produce a spatial pattern, a particular realization of the structure's operation. Many such realizations may be feasible (as suggested by information theory and entropy-maximizing methods; Chapter 4); only by understanding how the structure operates, and perhaps how the schema does too, can the causes of the particular pattern be comprehended, and its recurrence avoided.

Structuralism seeks to identify the global contexts of observed patterns, therefore, so that they are studied not as unique events with unique causes but as examples of the operation of a set of rules, which have produced a particular form at a particular time, perhaps because of a certain conjunction of constraints. Thus, as Thrift (1979) has argued, the 'inner-city problem' identified in Britain in the mid-1970s was not a unique event requiring its own interpretation, but rather the contemporary outcome of the operations of the capitalist mode of production. There are, however, few examples of this type of analysis in the geographical literature (McTaggart, 1974).

Marxism is seen by many as a form of structuralism, because it argues that spatial patterns—the superstructure—are realizations of a structure—the mode of production. To some, the cam-shaft analogy is invalid, however, for they interpret marxism as a form of economic determinism with no degrees of freedom in the production of the pattern. (Ley, 1978, for example, terms marxism a dogma based on certainty so as to protect itself.) Gregory (1978a) counters this argument with the concept of structuration, which is that 'man is obliged to appropriate his material universe in order to survive and because he is himself changed through changing the world around him in a continual and reciprocal process' (pp. 88–9). But structuration, on the other hand, does not argue that spatial pattern is a consequence of spatial structure, as indicated in Figure 6.1. Instead each requires the other: spatial patterns cannot be

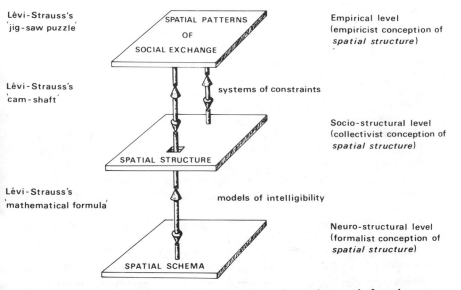

Figure 6.1   A geographical interpretation of Levi-Strauss's cam-shaft analogy for structuralism. Source: Gregory (1978a, p. 100).

theorized without reference to spatial structures, but also spatial structures cannot be operated without spatial patterns on which they are enacted (p. 121). Thus, for example, whilst the internal spatial pattern of city residential areas reflects the class system in the spatial structure, that system is itself composed, in part, as a consequence of the spatial patterns.

This two-way interaction between spatial patterns and spatial structures is an example of what is more generally known as the dialectical method. Developed by Hegel and adopted by Marx, this is summarized by the classic trilogy: thesis—antithesis—synthesis; an 'object' is opposed by its negative, and out of the conflict between the two a new 'object' is formed. Marchand (1978) has argued for application of this method in geography, because 'Anything that exists, undergoes perpetual change but maintains its identity at the same time. From this viewpoint, any real object is a centre of actions and reactions, and is a *subject*, although not necessarily a conscious one' (p. 108). Study of, for example, a city neighbourhood should therefore include its conflict with other areas in order to maintain its character, thereby through the conflict in some ways both changing itself and the other areas. City and country, centre and periphery are both examples of two areas in clear dialectical relationships, continually transforming each other. To study their mutual transformations is to get behind society's ideological smoke-screens and to destroy the myths of a value-free, positivist social science and its trivial empiricism.

Radical theory, research and action
By the mid-1970s, radical reinterpretations had been offered for several areas of geographical investigation, but, as Peet (1977) notes, most effort had been concentrated on the major topics of interest in the geography of the preceding paradigm. Many of these reinterpretations followed Harvey's dictum (p. 150) that further empirical work is unnecessary, and so their emphasis has been on theory.

The residential pattern of cities and the operation of housing markets has been a major focus of urban geographers' attention (Herbert and Johnston, 1978). A critique of the positivist approach to the study of those markets was presented by Gray (1975), highlighting:
1 the assumption that people are free to choose where they want to live, when in fact many people are forced to live in particular types of homes in particular places;
2 the assumption that residential patterns result from the amalgamation of a large number of decisions by individual householders, rather than by a few developers and financial institutions; and
3 'there is an implicit belief that the study of individuals and households provides the key to understanding urban processes' (p. 229).
Key words of the positivist paradigm, according to Gray, include aspiration, preference and choice, but:

There is an accumulating body of evidence from sociologists, and to a lesser extent from geographers . . . that people are not free to choose and prefer from a range of options, and that study of household units does not provide the key to understanding urban processes. Instead, many groups are constricted and constrained from choice and pushed into particular housing situations because of their position in the housing market, and by the individuals and institutions . . . controlling the operation of particular housing systems (p. 230).

To Gray, constraint and competition are more important concepts than are choice and preference, and

institutions rather than household units are the key to understanding urban processes . . . most fruitfully analysed as reflections of capitalist society. The structural analysis of capitalism and its various spatial manifestations is the core problem facing geographers (p. 232).

While agreeing with Gray that the research emphasizing preference and choice has been one-sided, Hamnett (1977) accused him of throwing the baby out with the bathwater and producing an equally one-sided analysis if he focused only on the production side of housing markets. One cannot study only the constraints and let the choices look after themselves, for many people do have to make choices, within a set of constraints. Gray, he argues, would be better to accept the findings of preference studies and build on them with a structural analysis, thereby producing a continuity in research, in the Popperian mould.

Urban residential patterns have been treated in a series of essays by Harvey, providing a structural analysis. Residential differentiation of various class groups produces community consciousness and the means of reproducing class differences; it is brought about by the actions of finance capital which are aimed at restricting choices and forcing up prices. The result is a series of spatially separate housing markets (Harvey, 1974e); within each, people may move around, expressing their preferences and thereby influencing micro-scale variations in the pattern.

But there is a scale of action at which the individual loses control of the social conditions of existence . . . [sensing] their own helplessness in the face of forces that do not appear amenable, under given institutions, even to collective political mechanisms of control (Harvey, 1975a, p. 368).

Among the processes at the scale beyond the control of the individual and the community collective is suburbanization, a major means of stimulating consumption—of housing, automobiles, consumer durables etc.—within the American economy at a time of dangerous latent overproduction (Harvey, 1975b, 1978).

Most of the radical work on housing markets and residential patterns in cities has followed Gray's rather than Harvey's point of view, focusing at the meso-scale on the managers who operate the processes involved in the production and allocation of housing. The concept of the 'urban manager' was developed by Pahl (1969), who argued that the residential pattern is the consequence of the operation of two basic processes:

1 There are fundamental *spatial* constraints on access to scarce urban resources and facilities. Such constraints are generally expressed in time/cost distance.

2 There are fundamental *social* constraints on access to scarce urban facilities. These reflect the distribution of power in society and are illustrated by:

bureaucratic rules and procedures

social gatekeepers who help to distribute the control urban resources (p. 215 in 1970 reprint).

The values and ideologies of these gatekeeper/managers are crucial to understanding the socio-spatial pattern of the city, an argument accepted by a group of workers at the University of Cambridge (e.g. Duncan, 1975), whose research included investigations of building-society lending policies (Boddy, 1976), the decisions of local authority housing managers (Gray, 1976), and the role of the local authority in various aspects of the private housing market (Duncan, 1974): similar work has been done in the United States on the operations of real-estate agents (Palm, 1979). Pahl (1975) later revised his ideas, pointing out that planners are merely the 'bailiffs and estate managers of capitalism, with very little power' (p. 7), so that focus on them and on other gatekeepers tends to

view the situation through the eyes of disadvantaged local populations and to attribute more control and responsibility to the local official than, say, local employers or the national government . . . such 'a criticism of local managers of the Caretaking Establishment' and 'of the vested interest and archaic methods of the middle dogs' may lead to an uncritical accommodation to the national elite and society's master institutions (pp. 267–8).

The managers are important elements, but are not themselves the spatial structure of Figure 6.1; they can be interpreted as the cam-shaft, operating the structure. Without care, treatment of the managers outside the constraints in which they operate could become as reductionist as aspects of behavioural geography (see also Williams, 1978).

Another major area of radical interest has been the modernization of the Third World, in the study of which (Drysdale and Watts, 1977) 'At times, geographers have resembled spectators who, seated on a hill overlooking a battlefield, are fascinated by the direction in which clouds of smoke are blowing' (p. 41). In the positivist mould, modernization had been investigated by Gould and his associates (e.g. Gould, 1970b; Riddell, 1970; Soja, 1968) who selected a series of variables representing various aspects of economic and social change (from a western, technological viewpoint) and manipulated these by multivariate statistical techniques to produce maps of the spread of modernization through colonial territories. Brookfield (1973) accepted that these provided useful descriptions of spatial patterns of change but argued that such 'value-loaded', although ostensibly 'value-free', research reveals little about the process of 'development': he characterizes the work as that of (Brookfield, 1975):

> A superb craftsman ignorant of the material with which he is working. The folly of such an approach was never better demonstrated than in the prolonged failure of . . . [Gould's] group to make its real contribution in an area where the direct participation of geographers is increasingly wanting (p. 116).

Even more worrying to Brookfield is the derivation of policies for development based on notions of diffusion within central-place frameworks. Berry (1972a) produced one of the main papers on this, one which, according to Brookfield (1975) is 'a highly mathematical paper on hierarchical diffusion employing his well-known, but doubtfully relevant data on the diffusion of TV in the United States' (p. 110). A further critique of the 'geography of modernization' studies (Slater, 1973) led Slater (1975) to identify what he saw as seven weaknesses of the 'Anglo-Saxon mainstream' abstracted empiricism approach:

1   its inverted methodology, which allows data to define problems rather than the reverse;
2   its surfeit of data relative to the power of its theory;
3   its mechanistic abstractions from socio-economic reality, which fail to account for such problems as under-development;
4   its concentration on form-description rather than on process-explanation;
5   its derivative and erroneous theories;
6   its failure to grasp the crucial role of the political economy; and
7   its lack of attention to the class structure and the role of class conflict in the structuring of space.

Brief examples are given of how a theory of uneven spatial development should be written, emphasizing the role of colonialism in the capitalist economy, and thereby moving the focus from the 'sterile perspectives' of the positivist work, highlighting instead that 'A crucial factor in the development of any spatial structure is the way in which surplus is circulated, concentrated and utilized in space' (p. 174). A similar attempt at theory-writing has been made by Folke (1973) with respect to imperialism.

Other aspects of geography for which a radical critique has been provided include Eliot Hurst's (1973) discussion of the impact of highway developments, studies of which have assumed that the socio-political environment in which the changes took place was itself unchanged. The impact of entrepreneurs on the transport system has been ignored, he claims; statistical procedures cannot account, for example, for the extortion rackets involved in the construction of California's railway network. For him:

> To really understand network evolution we need to know something about the decisions behind its construction. . . . To interpret this material, place it in a socio-economic and political framework, overcome the obstacles barring the way to the attainment of a better, more humane, more rational social order, and better explain and understand transportation as a factor in the landscape, should be our ultimate goals in transportation geography (p. 171).

In part, this appears to be a call for an idealist approach (p.132), except that Eliot Hurst mentions study of the political and economic frameworks as well as the actions of the individual decision-makers. The argument, according to Rimmer (1978), is the third phase in the development of transport geography, involving a repudiation of the preceding descriptive and quantitative/predictive phases. Its effect has been:

> to create confusion among transport geographers as the borrowings from sociology and politics have clashed with those from economics and mathematics. This confusion has created strain and tension as transport geographers have tried to adapt their ideas to a changing socio-economic environment distinguished by uncertainty and turbulence. . . . Transport geographers are, therefore, in a dilemma (p. 85).

As a consequence, Rimmer sees a need for a fourth phase, one of redirection, to produce a humanistic transport geography. Its framework

> has to encompass interactions between individuals and groups in such a way that it embraces distributional issues and equity goals in a spatio-temporal frame under either advanced capitalism or neocolonialism . . . transport policy has to be directed towards minimizing differences in personal mobility and reducing the polarization of opportunities produced by different methods of travel (pp. 80–1).

Such a redirection clearly does not fit the radical critique.

Finally, two examples can be quoted of work on environmental issues using a marxist framework. Harvey (1974d) has shown that contemporary statements on resource-population interrelationships are ideological, through an examination of the respective works of Malthus, Ricardo, and Marx. He argues that 'things' cannot be understood in isolation from their interactions with other 'things', so that resources can only be defined in the terms of the relevant mode of production. Thus, predictions such as those in the *Limits of Growth* study (p. 157) represent a status quo theoretical perspective (p. 150). Other viewpoints may lead to other conclusions:

> let us consider a simple sentence: 'Overpopulation arises because of the scarcity of resources available for meeting the subsistence needs of the mass of the population'. If we substitute our definitions (of subsistence, resources and

scarcity) into this sentence we get: 'There are too many people in the world because the particular ends we have in view (together with the form of social organization we have) and the materials available in nature, that we have the will and the way to use, are not sufficient to provide us with those things to which we are accustomed' (p. 272).

According to the first version of the sentence, the only policy option available is a reduction in population: according to the second, Harvey identifies three possibilities:

1   changing the desired ends and the social organization;
2   changing technical and cultural appraisals of nature; and
3   changing our views about the capitalist system and its scarcity basis.

The Ehrlich-Commoner arguments (p. 155) are based on the first version of the sentence and, according to Harvey, have been grasped by the elite of the capitalist world as the case for a popular ideology (the need for birth control) which will stabilize the capitalist system currently under severe economic stress.

Buchanan (1973) has argued in similar vein, contending that the neo-Malthusian case for birth control in the Third World is part of the 'white north's' imperialist policy of ensuring its continued access to the resources of the international periphery. The problems of ensuring a food supply for an expanding population, plus a supply of needed fuels and minerals, are not insurmountable, he claims, particularly if the residents of the Third World are granted the right to work. 'The poverty which is regarded as symptomatic of reckless population growth is rather a *structural poverty* caused by the irresponsible squandering of world resources by a small handful of nations' (p. 9). The 'white north' is operating an economy of waste, and its theories of overpopulation are grounded in that economy: the apparent humanitarianism of such self-seeking philosophies is, Buchanan argues, immensely dangerous.

Liberals and radicals in debate

The material discussed in this chapter illustrates very considerable differences of opinion among human geographers during the 1970s, differences which to many are indicative of an ongoing revolution, or paradigm-shift. They form the basis for a debate which, in its published components at least, is much more heated than that generated by the 'quantitative revolution' two decades earlier. It has also been developed in more depth and detail, using forums, such as the journal *Area*, which were not available twenty years earlier, when publication outlets for 'views and opinions' were few. Whereas the behaviourists caused little real concern within the discipline, and their views were soon coopted within the corpus of acceptable approaches, the radicals have had much more impact, because they attack not only the basis of most geographical work but also, in their clear inter-disciplinarity, the bureaucratic structure of the discipline and the existence of geography itself Johnston, 1978c).

A clear example of the liberal-radical polarization is given by a debate on the geography of crime. This was initiated by Peet (1975b), who argued that in attempting to make their work relevant geographers avoided asking 'relevant to whom?'; the political consequences of their work were ignored. The studies of crime reviewed (e.g. Harries, 1974) refer only to the surface manifestations of a social problem; they cannot provide solutions to that problem, but only ways of ameliorating it. 'So it is that "useful" geography comes to be of use only in preserving the existing order of things by diverting attention away from the deepest causes of social problems and towards the details of effect' (p. 277). Furthermore, geographers study only the crimes for which statistics are collected, thereby accepting the definition of crime used by the elite; the maps which they produce, which are useful to police patrols, can therefore be employed to help maintain the status quo of power relations within society. The position of the geographers involved, a position which they do not declare, is one of protecting the 'monopoly-capitalist' state.

Harries (1975) responded by attacking the simplistic nature of Peet's arguments, and claimed that geographers would have no influence at all if they argued merely that crime is a consequence of monopoly capitalism. He argued that it is best to work within the system, to make the administration of justice more humane and equitable, to protect the potential victims of crime, and to provide employment opportunities for graduate geographers. Approaches based on the control of crime are likely to be more influential than polemics relating its cause to the mode of production: as Lee (1975) also pointed out, Peet 'failed to provide us with any clues as to how he or other radical geographers would study crime' (p. 285).

Peet (1976a) responded to Lee's challenge, presenting a radical theory which would 'contribute directly, through persuasion, to the movement for social revolution' (p. 97). The theory argues that capitalism harnesses human competitive emotions and produces inequalities of material and power rewards. Aggression is an acceptable part of being competitive, and is often released on the lower classes, who are kept in the industrial reserve army, being encouraged to consume but provided with insufficient purchasing power. As the contradictions of this paradox increase, so does the pressure to turn to crime. Thus crime occurs where the lower classes live, and at the spatial interface between them and the middle class. Harries (1976) replied that being a radical was a luxury few academics could afford; working at a publicly-financed university demanded a pragmatic rather than a revolutionary approach unless unemployment was desired. To him, Peet's theory is overly economic and deterministic; it fails to account for cultural elements, such as the disproportionate criminal involvement of blacks, the sub-culture of violence in the southern United States and other areas, and the fact that all economic systems produce minorities disadvantaged in terms of what they want and what they can get by socially legitimate means. He could offer no alternative theory, however: 'I do not carry in my head a theory of crime causation, and I am quite incapable of synthesizing and attaching value judgements to existing theoretical formulations within a couple of pages of typescript' (p. 102). He encouraged Peet to come off the fence, and to get involved in the production of change within the present system. Wolf (1976), on the other hand, claimed that by concentrating on the traditional concerns of marxism Peet was not radical enough.

Two other geographers who have been involved in a considerable, often virulent, debate are Brian Berry and David Harvey. Berry (1972b) initiated the exchanges with his comments on Harvey's (1972) paper on revolutionary theory and the ghetto. Berry wondered whether Harvey's rational arguments on the need for a revolution would be accepted: 'because of "commitment", the opposition will quietly drift into corners, the world will welcome the new Messiah, and social change will somehow, magically, transpire' (p. 32). The power to achieve change, he contends, needs more than logical argument to produce it in the twentieth century—'nothing less than cudgels has been effective' (p. 32)—and Harvey's belief in logical rationalism will be to no avail. He has also argued that, in any case, Harvey is wrong about the ghetto, for liberal policies are succeeding and the inequalities between blacks and whites are being reduced (Berry, 1974a). Harvey (1974a) responded that scarcity must continue in a capitalist economy, which will leave some people—those in the inner city—relatively disadvantaged.

In reviewing *Social Justice and the City* (Harvey, 1973), Berry (1974b) criticized Harvey's dependence on economic explanations. Basing his case on the arguments of Daniel Bell (1973) on post-industrial society, Berry (1974b) claimed that the economic function is now subordinate to the political: 'the autonomy of the economic order (and the power of the men who run it) is coming to an end, and new and varied, but different, control systems are emerging. In sum, the control of society is no longer primarily economic but political, (p. 144). Harvey's (1974b) response was that marxism could not be considered as passée while the selling of labour power and the collusion between the economically and the politically powerful all continue, and that the state has to be considered too within a marxist framework (Harvey, 1976). Berry (1974b) retorted:

> I believe that change can be produced *within* 'the system'. Harvey believes that it will come from sources *external* to that system, and then only if enough noise is produced at the wailing wall. . . . The choice, after all, is not that hard: between pragmatic pursuit of what is attainable and revolutionary romanticism, between realism and the heady perfumes of flower power (p. 148).

Harvey's (1975c) review of Berry's (1973a) *The Human Consequences of Urbanization*—a study of urbanization processes at various times and places and of the planning response to these—concluded that the book is 'all fanfare and no substance' (p. 99), revealing that Anglo-American urban theory is substantively bankrupt and that 'it is scholarship of the Brian Berry sort which typically produces such messes' (p. 99).

> It is doubtful if it makes any sense even to consider urbanization as something isolated from processes of capital formation, foreign and domestic trade, international money flows, and the like, for in a fundamental sense urbanization is economic growth and capital accumulation—and the latter processes are clearly global in their compass (p. 102).

To Harvey, Berry has nothing to say of any substance, but as Berry 'is influential and important . . . his influence is potentially devastating' (p. 103). Berry's only response was a general comment on the Union of Socialist Geographers (Halvorson and Stave, 1978):

> there's no more amusing thing than goading a series of malcontents and kooks and freaks and dropouts and so on, which is after all what that group mainly consists of. There are very few scholars in the group (p. 233).

Apart from these very polarized exchanges, there have been several other statements which indicate that whereas some 'liberals' have been pre- pared to consider the radical case seriously, others have tended to avoid the issues, more or less. Chisholm, for example, has (1975) claimed that:

> while I am fully sympathetic to the view that the 'scientific' paradigm is not adequate to all our needs, and must be supplemented by other approaches, I am not persuaded that it should be replaced. . . . Harvey wants us to embrace the marxist method of 'dialectic'. This 'method' passes my under- standing; so far as it has a value, it seems to be as a metaphysical belief system and not—as its protagonists proclaim—a mode of rational argument (p. 175)

(see also Chisholm, 1976). More frequently, reviewers accept that the radical view is valuable but not, to them, tenable in its entirety. Thus Morrill (1974) writes of Harvey's *Social Justice and the City* that 'I am pulled most of the way by this revolutionary analysis but I cannot make the final leap that our task is no longer to find truth, but to create and accept a particular truth' (p. 477). King (1976), too, seeks a middle course,

> An economic and urban geography that will be concerned explicitly with social change and policy. . . . Such a middle course will not find favour with the ideologues, who will see it either as another obfuscation favouring only 'status quo' and 'counter-revolutionary' theory, or as a distraction from the immediate task of building elegant quantitative-theoretic structures, but some paths are being cut through the thicket of competing epistemologies, rambling lines of empirical analysis, and gnarled branches of applied studies that now cover the middle ground (pp. 294–5).

He accepts that much quantitative-cum-theoretical geography has sought mathematical elegance as an end in itself, at the sacrifice of realism; he believes that social science must feed into social policy and generate social change; he accepts the 'intellectual power' of Marxist analysis but believes that the prescriptions based on it are acceptable only if the ideological framework is: his conclusion suggests the need for more quantification, aimed at being operationally useful rather than mathematically elegant. Such points are extended in a later paper (King and Clark, 1978), which concludes both that 'the state must be explicitly treated as an element to be integrated with the more orthodox models of regional development' (p. 11) and that '*space* . . . should be seen as an element in the political process, an object of competition and conflict between interest groups and different classes' (p. 12). And finally, Smith (1977) concluded that:

> Marx may have been able to dissect the operation of a capitalist economy with particular clarity, and see the essential unity of economy, polity and society that we so often miss today. But Marx does not hold the key to every modern problem in complex, pluralistic society (p. 368).

Conclusions

This chapter has brought the debate on the nature of human geography up to date at the time of writing, and has indicated the depth and breadth of the discussions which have occupied much of the 1970s. The clearest conclusion which can be drawn is that there is an increasing proportion of geographers who wish to be involved in the re-shaping of local and world societies, either through ameliorative correction of current problems and trends or by designing desirable spatial organizations (Berry, 1973a). Their motives range along the continuum from 'pure altruism' to 'devoted self-seeking'. Their methods vary from those who accept the present mode of production and see humanitarian goals as achievable within its constraints, through those who subscribe to the phenomenological view (Buttimer, 1974) that:

> the social scientist's role is neither to choose or decide for people, nor even to formulate the alternatives for choice but rather, through the models of his discipline, to enlarge their horizons of consciousness to the point where both the articulation of alternatives and the choice of direction could be theirs (p. 29),

and ending with those who believe that a revolution is necessary to remove the causes of society's myriad problems and replace them by an equitable social structure.

Perhaps somewhat surprisingly, the 'revolutionaries' of the above classification are not the current 'activists'. It is the liberals who believe that geographers can contribute to the immediate solution (or amelioration) of current problems and who are pressing for academic involvement in policy-making: the radicals have a longer-term goal of education which will generate a mass movement demanding revolutionary change. The two groups are members of the same discipline. Contact between them is either virulent or non-existent, it would seem from their published exchanges. And alongside them are those little affected by the debate, whose interests, be they in historical, cultural or regional geography, lead them to believe that their task is to continue as before, sensitizing their readers to the world as they, the geographers, perceive it.

7

Evaluation

The preceding five chapters have presented the substance of this book, outlining the major trends in Anglo-American human geography during a period of about three decades. The chapter titles and contents refer to different approaches to human geography, not to periods of years, and a general concordance of the two has yet to be accounted for. Thus this final chapter returns to the subject matter of the Introduction (Chapter 1). Having described the changing content of human geography, the aim now is to try and understand these changes, why they occurred, when they did (little attention is paid to *where* they did, though this would undoubtedly make an interesting topic for study). Some suggested reasons for the changes have been outlined in the body of the book; an attempt is made to integrate them here.

Although it has only occasionally been referred to during the preceding 174 pages, it will be clear that the paradigm model introduced in Chapter 1 (p. 15ff.) has provided the major organizational guideline and format for the whole presentation. It is an evaluation of the validity of that model which occupies the present chapter: as indicated in the Preface, no attempt is made in the book to assess either the contributions of geographers as a whole or the work of particular geographers to the body of generally available knowledge. The concern of the present book is what was done, and why, not how it was done and what effect it had.

The paradigm model has been chosen for two reasons. The first is its attraction as a plausible, coherent descriptive device. The second, and probably the more important, is that it has been adopted by many others for the purpose of accounting for changing scientific tastes; those using it within human geography include Haggett and Chorley (1967), Berry (1973b), Harvey (1973), Herbert and Johnston (1978) and Buttimer (1978b). Not all geographers have been attracted to the model, however, even as a simple descriptive device. Bird (1977), for example, has claimed that if Taylor (1976) could identify seven revolutions during the period currently under review—the seven are quantitative, methodological, conceptual, statistical, models, behavioural, and radical—then, 'Perhaps so many revolutions in so short a time indicate in themselves either a continuously rolling programme, or something basically wrong with the overturning metaphor' (p. 105). A year later he (Bird, 1978) entitled a section of his paper 'Paradigm as "the one and only" is dead' (p. 134), claiming that 'Monistic methodological solutions for social science seem to be falling out of fashion, perhaps reflecting the fact that society itself is organized around more than one major principle' (p. 134). Even more stringent criticism of the paradigm model and its relevance to the history of geography has been presented by Stoddart (1977), who enquires 'Whether the paradigm idea is useful in understanding processes of change in geography on other than a superficial level' (p. 1). To him the answer is no, for:

> the concept sheds no light on the processes of scientific change, and readily becomes caricature. I suggest that as more is understood of the complexities of change in geography over the last hundred years, and especially of the subtle interrelationships of geographers themselves, the less appropriate the concept of the paradigm becomes (p. 1).

Indeed, he argues that

> There is scope for sociological enquiry into the extent to which the concept has been used in recent years as a slogan in interactions between different age-groups, schools of thought, and centres of learning (p. 2).

Such remarks provide the context for the evaluation in the next section.

The paradigm model evaluated

Superficially, as Stoddart suggests, the paradigm model would appear to fit what has happened in Anglo-American human geography since the Second World War; this is clearly implied by the organization of the present book.

Until the 1950s the regional concept and a regional approach dominated the endeavours of researchers and teachers, although, as the differences between Hartshorne and Sauer indicate (Guelke, 1977b), there was no homogeneous conventional wisdom. But then, as Gould (1975) expresses it, a new type of work was initiated

> by geographers with quite conventional backgrounds and training, by men who had made strong professional commitments to their chosen field, but who were distressed by the lack of true intellectual demands characterizing much research during the preceding decades (p. 305).

This was the positivist movement, described in Chapter 3, with its multiple origins in the United States: it developed into what has been described in Chapter 4 as spatial science.

Over a decade or so, spatial science progressed in the manner described by Kuhnian 'normal science', sharpening its tools, testing and revising its theories, and extending its substantive sphere of influence. A significant challenge to its status as orthodoxy came in the mid-1960s with a behaviourist approach that criticized what were seen as weaknesses in the axiomatic underpinnings of its theories. This challenge was met by incorporation and further modification of spatial-science theories, and the paradigm was only slightly deflected.

By the late 1960s, however, a further challenge was developing. This was a behavioural approach, with a strong humanistic flavour, an apparently reductionist philosophy, and roots in the earlier, but by then widely discredited cultural, historical and regional geography. But no revolution occurred. The challenge was not defeated, however, nor did it lead to the establishment of a new sub-discipline: instead, according to the bureaucratic procedures discussed in Chapter 1 (p. 21ff.), it became established as a revolutionary enclave within the geographical corpus, making a few converts but not expanding substantially.

A further attempted revolution was launched in the early 1970s, as the spatial scientists attempted to apply their expertise to solve problems in the economic and social structures of western (and, to a lesser extent, Third World) society. Structuralists, notably marxists, argued that spatial science provides only a surficial analysis of the spatial realizations of socio-economic processes, an analysis which by its methods cannot penetrate, let alone comprehend, those processes; without such comprehension, social and economic change cannot be achieved, so spatial science is of no value in improving the average quality of life in the world. Thus the structuralists aimed at a revolution which would replace spatial science by something more realistic and capable of application. They achieved no immediate success, however, despite a number of conversions, and, as Chapter 6 illustrates, their debate with their adversaries defending the orthodoxy made a considerable impact on the literature of the mid and late 1970s.

Outwardly, therefore, human geography appears to have experienced one complete revolution, during the period studied here, in the 1950s and early 1960s. Another was attempted at the end of the 1960s, but although a 'revolutionary party' was established and maintained a foothold, it did not succeed in its attempt to dislodge the entrenched orthodoxy. A third attempt at a coup evolved during the 1970s, but again, although this structuralist approach was probably more successful than its predecessor, it did not manage to overthrow the hegemony of spatial science. These episodes can all be associated with outside events, as described in Chapter 1 (p. 23ff.). The first revolution was clearly related to the era of science and planning which dominated the years following the Second World War and the ensuing conflicts in Korea, Vietnam, and elsewhere. The behavioural moves within geography were not stimulated by major trends elsewhere in society, although by the late 1960s science and planning were increasingly under attack both for their failures and for their neglect of the welfare of the individual *qua* individual. And finally, the structuralist attacks of the most recent decade represent the wider demands for sweeping social and economic change which have attended the faltering of 'western' economic growth during this time (Peet, 1977).

Clearly, then, there have been shifts of emphasis within human geography, and important debates as to the nature of, and need for, such shifts. But have there been any true revolutions? Have geographers, both collectively and individually, travelled down the Damascus road several times in only a few years? The evidence is far from convincing. The title of Gould's (1975) paper, for example, asks 'Mathematics in geography: conceptual revolution or new tool?'—although he does not essay an answer. And even a casual glance at the major periodical literature of the late 1970s will indicate that much research is being done in all three of the 'paradigms' outlined above. Indeed, cases have been made that such eclecticism is to be encouraged (e.g. Wilson, 1978) and there are clear signs that spatial science and systems analysis were still widely practised (on the former, see Getis and Boots, 1978 and Haggett, Cliff and Frey, 1977; on the latter see Chapman, 1977 and Bennett and Chorley, 1978).

Several paradigms were advocated for geographers' attention and acceptance during the 1970s, therefore; each could be subdivided into separate, if not independent, sub-paradigms. While the decade may be one of revolutionary struggle, it could not be characterized as one of revolution. Was the same true of the 1950s? As indicated in Chapter 3, many of the 'revolutionaries' then active stressed the continuity of their interests with those of their predecessors, and Guelke (1977b) has claimed that:

> To an extent that is not widely recognized, the move to quantification took place within the basic framework of geography put forward by Hartshorne in *The Nature of Geography*. For the vast majority of North American geographers . . . the replacement of map-overlay techniques of correlation with more accurate statistical and mathematical procedures involved no change in basic philosophy . . . [Hartshorne's] widely diffused ideas were essentially well disposed to the adoption of the new methods (p. 3).

As spatial science took over, however, so

> the new geographers abandoned the idea of geographical significance in its traditional meaning and, at the same time, eliminated the need for regional geography as the central core of the discipline (p. 4).

The acceptance of a new central core implies the succession of a new paradigm, as suggested also by Chisholm (1975):

> Although perhaps only a change in emphasis, the magnitude of the alteration is sufficient in itself to constitute a revolution, a revolution that could be described as placing description in the role of handmaiden to explanation (p. 171).

But, he argues, although approach may have changed, in effect the trends of the 1950s and 1960s reflected a move back to the concerns of the pre-regionalism period: 'Therefore, in respect of analytical techniques geography has indeed experienced a revolution. However, in terms of subject-matter the position has changed less rapidly' (p. 173). So much, then, for the 'quantitative and theoretical revolutions'. As for the 1970s, perhaps this would be seen as the decade 'in which the problems geographers discuss caught up with the analytical skills the profession can now boast' (p. 174).

Clearly, then, there are contradictions, and not all commentators would recognize the period since World War II as one of several major revolutions in the nature of geographical thought. To establish with more clarity what has occurred, therefore, it is necessary to return to a consideration of the paradigm model itself.

The paradigm model revisited
In his discussion of the paradigm model and its relevance to geography, Stoddart (1977) argues that 'This ready acceptance of Kuhn's vocabulary has occurred without any close attention to Kuhn's own statements or to the critical literature on them' (p. 1). This is indeed so, for the bibliographies of most of the geographers who base their exegesis on Kuhn's model contain few, if any, references to the literature which it has generated. Most geographers apparently rely on the first edition only (Kuhn, 1962): they seem to be unaware that (Suppe, 1977a) 'The loose use of "paradigm" in his book has made *The Structure of Scientific Revolutions* amenable to a wide variety of incompatible interpretations' (p. 137) or that (Suppe, 1977c) 'Kuhn's views have undergone a sharply declining influence on contemporary philosophy of science' (p. 647). Indeed, Kuhn (1977) has reinterpreted his position several times since the original formulation was produced.

A review of the whole of the complex debate on paradigms and normal science would be out of place here. There are clearly many obvious disagreements among philosophers/historians of science on the subject. Watkins (1970), for example, argues that Kuhn's formulation of the 'paradigm-switch mechanism' (the nature of revolutions) suggests irrational behaviour both by the scientists defending the old paradigm ('narrow-minded dogmatists') and by those propagating the new one ('religious fanatics'). Stegmuller (1978) has disputed this, however, and analyses the nature of revolutions in great detail. Nevertheless, the preceding section of this chapter has illustrated Watkins's contention (p. 19) that sciences should be in a constant state of dual-paradigm, if not multi-paradigm, debate and Harvey's (1973) query as to what constitutes a revolution (see p. 159).

An important topic within the general debate that is of relevance here concerns the degree of consensus in a science. Is it reasonable to expect an entire discipline, especially one with more than a thousand active researchers, to be characterized by a single paradigm, by a large body of scholars pursuing exactly the same ends with exactly the same means, and not differing at all, except occasionally (when a revolution occurs), on fundamental interpretations? Could this be expected of any group, let alone a body of scholars dealing with material as complex as modern capitalist societies? Mulkay's (1978) evidence, from his research on radio astronomy, indicates that 'Scientific consensus . . . in a given area of interest is seldom complete' (p. 111), and when it does come about, it is because many of the conflicts are internalized within the discipline as a result of mutual accommodations to maintain disciplinary status (see p. 21). Even so, 'there are now several well documented studies of instances where demonstrably competent scientists have been excluded from a field of study as their ideas have come to diverge from those of the majority' (p. 116). Kuhn (1977) accepts this, noting that 'Individual scientists, particularly the ablest, will belong to several such groups (i.e. research communities), either simultaneously or in succession' (p. 462), and Suppe (1977b) points out that, as individuals, all scientists will have their particular viewpoints and will place varying stress on the ideas of competing groups. To him, however, for a researcher

> to account for normal science all he has to assume is that scientists in a particular scientific community are in sufficient agreement on what theory to employ, what counts as good and bad science, what the relevant questions are, what sorts of work to take as exemplary, and so forth, and communicate freely enough that only certain characteristic problems are seriously considered. To account for revolutions, one only needs to say that such unanimity breaks down in certain characteristic ways, and that their breakdowns typically get resolved in certain ways (p. 498).

A major element in this discussion of consensus would seem to refer to the scale of analysis. Two separate scales appear appropriate. The first is that described in the opening sentence from the above quotation by Suppe, and refers to a general acceptance of what are acceptable as ends and means. This is what Kuhn (1977) terms the paradigm as a disciplinary matrix: 'what the members of a scientific community, and they alone, share . . . it is their possession of a common paradigm that constitutes a scientific community of a group of otherwise disparate men' (p. 460). In contrast to this world view, at the second scale a paradigm can be defined as a series of applications of the disciplinary matrix. In this case, Kuhn terms it an exemplar, a 'model' example of the application of the matrix to a particular type of problem. A training in a certain science, therefore, involves acquiring an 'arsenal of exemplars' (p. 471) relating to its range of applications. Once trained in the broad field, the individual scientist may then join one or more of its specialist groups, each of which is focused on a particular type of application, with its own exemplars, and probably also one or a few major research leaders.

This resolution of the paradigm into two scales accommodates Mulkay's (1975) notion of research branches within an established discipline (see p. 21). Different applications of the same disciplinary matrix frequently develop, with relatively little contact between them, although an active scientist may participate in more than one simultaneously (despite the problems of keeping abreast of a burgeoning literature): senior members of the discipline will act as assessors and patrons in the evaluation procedures for appointing new academics and promoting others (see p. 11). Thus within a paradigm, the various branches may promote differences which, in time, could generate dissension and perhaps lead to revolutionary situations as the work in one exemplar group leads away from the general philosophy of the disciplinary matrix. A 'quiet revolution' may be the outcome, and the sort of debate identified in Chapter 6 will occur only when either different exemplar groups compete for disciplinary hegemony or when 'true' Kuhnian revolutions are fostered.

Eclecticism is possible within a disciplinary matrix, therefore, perhaps more so in a social than in a physical science. Different groups (schools of thought?) are able to coexist, sharing the same general philosophy but pursuing somewhat separate goals; each recognizes the quality of work in the others. The relative importance of groups will vary over time, perhaps occasionally leading to a 'quiet' revolution in a new set of ideas becomes dominant and creates an alternative disciplinary matrix.

Disciplinary matrices, exemplars, and human geography
In applying this reinterpretation of the paradigm model to recent human geography, it is necessary at the outset to decide whether the period under consideration was preceded by one in which a single paradigm dominated: the alternative is to conclude that human geography has been in a permanent pre-paradigmatic state (p. 18), with frequent reformulations and contests for 'conventional wisdom' status, but no constant body of accepted knowledge.

It would appear that the era dominated by regionalism had a clear disciplinary matrix, much of it codified by Hartshorne. Geographical activities involved the application and reapplication of the arsenal of exemplars to different areas, and there was development of new tools, some successful, as with Sauer's 'Berkeley school', and some less widely appreciated and adopted, such as Whittlesey's compage concept. And then, as Guelke (1977b) describes it, the quantitative revolution began. According to both the quantifiers and the above reinterpretation of the paradigm model, this revolution began as a new branch of the regionalism disciplinary matrix, accepting the general goals of geographical scholarship but developing its own sets of exemplars, very different from those employed elsewhere. There was some debate on the merits of these new exemplars—whether, in Suppe's terms (p. 180 above), they qualified as 'good geography'—but Burton (1963) suggests that they were soon, if grudgingly, accepted by the establishment of the discipline (see also

Taylor, 1976), perhaps because they had charisma, attracted students and outside attention, and promised to contribute to the growth of geography at a crucial period of expansion in academic research and teaching.

The break was a slow one, as the 'quantitative approach to areal differentiation' changed into 'spatial science'/'locational analysis'. A new disciplinary matrix was established, and a revolution seemed to have occurred, somewhat by stealth, from within. Much regional geography was now patronized as 'mere description', and failed to meet the canons of the new science.

Was (is) spatial science a paradigm? Apparently, yes. A disciplinary matrix was assembled, and eventually codified by Harvey (1969a), while early texts, such as Haggett's (1965c), provided major exemplars—as did the papers of Garrison, Berry, McCarty and others. The usual processes of normal science then operated, with the range of applications and the arsenal of tools being extended, thereby providing new exemplars. Comparison of the first and second editions of *Locational Analysis in Human Geography* (Haggett, 1965c; Haggett, Cliff and Frey, 1977) indicates the extent of this normal scientific activity and illustrates the nature of progress within the paradigm (Wise, 1977).

Within the spatial-science disciplinary matrix there was soon a proliferation of branches, representing groups of workers who applied the accepted positivist method and statistical procedures in particular ways to particular subject matters. Some of these branches were, as van den Daele and Weingart (1976) outline in a general review, reductionist, involving the splitting of existing branches. (An excellent example is provided by urban geography, which by 1970 could be divided into two obvious branches—the 'central-place analysts' and the 'social-area analysts': most urban researchers were active in one of these only.) Others were aggregative, combining geographical interests with those from other disciplines (as, for example, in some of the work in medical geography): yet others were integrative, as in the growing interest in environmental issues. The contact between the branches was often slight—within the Institute of British Geographers, for example, a number of separate study groups was formed (each member could belong to two without paying a further subscription) and these now operate both separate conferences and their own sessions at the Institute's annual conference. Perhaps the separateness is best exemplified by Gould's (1972) belief in the need for a pedagogic review of Wilson's (1970) work on entropy-maximizing modelling.

One of the branches which developed in this was the behaviourist one identified in the first half of Chapter 5. This was absorbed within the main body, and did not threaten even a 'quiet' revolution because it accepted most of the canons of the spatial-science disciplinary matrix. Indeed, it soon developed its own branches—in the study of voting, for example, and in industrial location research. The behavioural work discussed in the second half of Chapter 5 did not develop as a branch of the existing disciplinary matrix, however. Instead, it grew as a revitalization of the earlier branches of regionalism (with the associated cultural and historical geography), to which new philosophies, such as phenomenology and idealism, were added (Parsons, 1977). From the outset, therefore, it was in a state of conflict with the existing disciplinary matrix, and yet, as suggested in earlier chapters (see p. 140), it lacked a clear disciplinary matrix on which to base its competition: its early statements were more negative (anti-positivism) than positive in tone.

The structural approaches, on the other hand, have begun much as spatial science did, as a branch of the orthodoxy, from which they became dissociated with the development of a revolutionary disciplinary matrix. Some of the early work in this genre—the radical approach—accepted much of the spatial-science disciplinary matrix, as Harvey (1973) and Peet (1977) indicate, and applied its exemplars in slightly different ways to a new constellation of topics, thereby creating a set of extra branches. Some of these still exist, such as the 'welfare geography' espoused by Smith (1977). In the early 1970s, however, as their disenchantment with spatial science grew, the structuralists broke away and began the revolutionary debate chronicled in Chapter 6.

Why, then, was the spatial-science revolution, when it came, such a conspicuous success; why has the behavioural effort not gained a major following; and is the structural breakaway likely to result in a revolution? The answers to these questions, it would seem, lie in a combination of factors, some internal to the discipline of human geography and others relating to its external, societal environment. On the latter, the pro-science attitudes of the immediate post-war years have already been discussed. To be scientific and to have a contribution to make to planning was to be reputable, and geographers soon realized that they could make a mark by offering to provide 'operational research in space': a collection edited by Daysh (1949) indicated an early appreciation of the potentials by British geographers, albeit not spatial scientists in their technical arsenal. The structuralist trend, too, reflects the wider societal matrix in which it emerged, with the economic and social difficulties of the 1970s and widespread perception of the failure of 'liberal capitalism'. The behavioural movement, on the other hand, had a less obvious base in society and its contemporary needs; although there were clearly links between its humanistic aims and the growth of the social sectors of the welfare states, it was the liberal positivism of 'welfare geography' which capitalized on perceived needs rather than the more esoteric and difficult, at least to the outsider, hermeneutic approaches.

Changes in the external environment are clearly relevant to changes in human geography, therefore; they provide a necessary, but probably not a sufficient condition for the launching of a successful revolutionary putsch. Associated with such changes must be a set of sympathetic conditions within the discipline itself. Several historians of science have noted, for example, that new paradigms are usually initiated by younger workers. Thus, according to Stegmuller (1977)

> it is mostly young people who bring new paradigms into the world. And it is young people who are most inclined to champion new causes with religious fervour, to thump the propaganda drums (p. 148),

and, according to Lemaine *et al*. (1976)

> Mendel's work, and that of his successors, was a response to scientific problems. But the scientific implications of their results were not pursued until there existed a strong group of scientists who, owing to their academic background and their position in the research community, were willing to abandon established conceptions (p. 5).

Such groups of young workers, reacting, in the case of social sciences such as human geography, to the demands of their constraining society, usually require the catalyst of one or more established iconoclasts able to lead branches.

Establishment of a new branch requires several criteria to be met (van den Daele and Weingart, 1976). These include: an autonomous system of evaluation and reputation; an autonomous communication system; acknowledgement of their ability to solve puzzles within the confines of the disciplinary matrix; a formal organization providing training programmes which allow reproduction and expansion of branch membership; an informal structure with leaders; and resources for research. Most of these are much more easily obtained at a time of general educational expansion, when both the establishment of new research institutes, or of groups within existing organizations, and the provision of funds for undergraduate and postgraduate training are generous. The late 1950s and much of the 1960s was such a period, and it allowed the spatial scientists to expand their numbers rapidly and achieve their dominance of the discipline, which now extends into the heart of its establishment. Since then, however, resources have been much less easily obtained. The structuralists have not wanted for either leaders or potential disciples, but relatively few of either category have obtained the permanent status within the disciplinary organization which could guarantee them access to needed resources: in part this reflects not only the shortage of resources but also lack of sympathy within the orthodox (i.e. spatial science) establishment. The behavioural group perhaps lacked the iconoclasts: with a few exceptions, their leaders have been individual scholars who have not generated large research followings and have not sought to dominate the resource-allocation procedures. Further, the external environment, dominated by positivist approaches in both the private and the public sectors of economy and society, has not been sympathetic to humanist suggestions for societal improvements.

What is being suggested, therefore, is a generational model of disciplinary development (Johnston, 1978c). The external environment creates the conditions which are favourable for a redirection of scholarly effort. An iconoclast (sometimes several), usually with secure position in the academic career system, reacts to these conditions by creating new exemplars, most often within the established disciplinary matrix, and attempts to obtain the resources which will allow initiation of research programmes based on his ideas. If successful, his followers may form a branch, or constellation of branches, which either comes to dominate the disciplinary matrix or, by revolution, replaces it. Clearly, then, the spatial-science development took place when all of the criteria were met: iconoclasts were available to compete for the large amounts of available resources and meet the new demands of a planning-oriented, science-based, mixed society. The behavioural and the structuralist groups have not been as fortunate and, as a consequence, the many appointments of young spatial scientists made in the 1960s have allowed their disciplinary matrix to establish a considerable dominance over human geography which conditions since have protected somewhat from further revolutions, quiet or otherwise.

One component of this generational model is clearly at variance with Kuhn's paradigm model: the latter postulates a major conversion of scientists from one paradigm to another at the moment of revolution, but the generational model assumes that most scientists continue to work in the paradigm to which they were socialized academically. Conversion from one paradigm to another is comparatively rare, it is argued here: human geographers may migrate from one branch to another within the same disciplinary matrix, but they are unlikely to make the major change that a revolution requires. (This is perhaps particularly the case with the structural/spatial science debate of the 1970s, since this has ideological components and would involve many spatial scientists abandoning their established political as well as their disciplinary matrices.) Thus, according to this model, a revolution only occurs within a discipline when a new generation of scientists, with its own philosophy and methodology, becomes numerically dominant.

The generational model evaluated
This generational model accounts for the variations between paradigms and branches in the volume of reported research, and thus their relative dominance of the discipline's journals. It is, however, very much focused on the research literature and begs certain questions, such as 'if spatial science became so dominant by the late 1960s, why were so many academic geographers apparently unable either to teach or to examine in the quantitative area during the early 1970s?'. The answers relate to the unrepresentative character of the research literature as an indicator of the educational content of the discipline. Law (1976), for example, has counselled caution in the interpretation of the research record:

> In practice it may well be the case that scientists do lay special emphasis on the accounts in scientific papers, but my hunch is that there is immense (and non-trivial) variation between scientists on this count. For some, science is something you do in the laboratory, something you talk about, and something you get excited about. For others, science is what they write and what they read in the journals. I would even hypothesize (in conformity with the invisible college notion) that those who are generally felt to be of high status locate science less in the journals than in their own and other people's heads (p. 228).

If this is the case in human geography, then not all branches extant within the profession need be reflected in the contemporary research materials. It is a common belief within academic life that 'unpublished research isn't research' and publication, as indicated in Chapter 1, is a major criterion of academic attainment. Nevertheless, many academics publish very rarely, if ever, in the journals containing their discipline's referred research reports.

Important in this context is the academic career cycle. For most members of the academic profession, there is a rapid decline in research activity and publication after the age of 40 or thereabouts, as they penetrate the higher levels of academic bureaucracy and channel their energies in other directions. As 'normal science' continues in their relative absence, so they become isolated from its developments. And yet the great majority of them continue to teach students—especially undergraduates—and most use the disciplinary matrix and exemplars which formed the basis of their own academic socialization. Thus the range and variety of courses taught in a university department may be much greater than is apparent from the research publications of its staff.

The effect of this recognition of the academic-career cycle would seem to be an extension of the generational model to indicate that the current research paradigms do not always reflect the variety of disciplinary matrices and exemplars taught: such a situation will be most likely when the nature of research activity is changing rapidly. The generational model requires broadening, and in this way a further feature of 'progress in geography' can be accounted for. Freeman (1961) points out that he was tempted to give his book *A Hundred Years of Geography* the subtitle ' "no new idea under the sun" for many ideas are produced, ignored, and revived fifty years or so later with good results' (p. 10). This reflects the continued teaching of outmoded, in a research sense, disciplinary matrices, either by the original iconoclasts or by their disciples. In a period of rapid change, many branches of different paradigms remain alive; occasionally they may be taken up by energetic workers, and used as bases for attacks on the current orthodoxy—which are taken up may be the result of a chance encounter between a good teacher and a student who is a potential iconoclast. This does not, of course, mean that new ideas are never introduced; the interaction between disciplines ensures that they are, often producing modifications of the existing exemplars.

But it is not only the continuity of teaching activity, relative to that of research, which can lead to new branches being established, with or without immediate paradigmatic ambitions. Popper (1967) has introduced the concept of three separate, though interdependent, worlds:

World One—the physical world of objective knowledge (this corresponds with Kirk's concept of the phenomenal environment—p. 126);
World Two—the subjective world of conscious experiences (Kirk's behavioural environment); and
World Three—the objective knowledge which is stored in books, computers, and like objects, which themselves exist in World One.

As Bird (1975) points out, geographical publications, along with those of all other researchers, are lodged in World Three, where they are available for future generations of researchers, whether or not teachers direct them towards such material. Thus literature searches may unearth ideas to be revived, as with the anarchist notions of a nineteenth-century geographer, Prince Kropotkin (Stoddart, 1975b). Branches, then, may never die, but merely fall dormant: their revival, of course, may be the result of a chance discovery, perhaps coupled with a belief in their relevance to an attack on the currently popular disciplinary matrix.

So are there geographical paradigms?
Is, then, the concept of a paradigm, of an agreed disciplinary matrix which dominates activity at any one time, relevant to modern human geography, or has it been a descriptive red herring? The evidence presented throughout this book, and particularly in the present chapter, favours the latter interpretation (as indeed might a similar study of earlier periods). Clearly, there have been several schools of thought active in human geography during the period studied. Most have been initiated by iconoclasts (to continue the use of Eilon's terminology): some of these obtained their stimulus from the earlier geographical literature; some incorporated ideas and methodologies that they had discovered in cognate disciplines; and a few may have had very productive ideas of their own. Many of the schools which they founded were only minor variations on others, attracted a few adherents, reported a body of research, then faded. Only a few, in favourable circumstances within the discipline, with charismatic leadership, and, perhaps most important of all, apparent relevance to the contemporary requirements of society, have expanded their spheres of influence rapidly, and have come to occupy, at least for a substantial period of time, a major niche in the discipline and perhaps dominance in its research literature. They have generated their own branches, and dominated the disciplinary matrices passed on to one or more generation of trainee geographers. But all are ephemeral, though resurrection is possible.

Why should this be the case? Why is human geography, and probably other social sciences, unlike the physical sciences in the relevance of the paradigm concept? All sciences have the same general goal, of understanding some aspect of the 'real world', but although the subject matter of the physical sciences may be no less complex than that of the social sciences, its procedures can be more precise, because of an ability to experiment and to isolate the component parts of the universe under consideration. Progress in physical science is towards clearly stated proximate ends, therefore; movement towards them can be monitored, and ideas discarded when they prove fruitless, because, in Popperian terms, they can be falsified. The social sciences, perhaps only because of their immaturity, cannot be so precise, and few ideas can be comprehensively falsified. Progress in geography, according to Wise (1977), should be part of the more general progress of mankind:

> The main condition must be the continued determination, effort and skill of human beings, through reasoned endeavour, to attain higher levels of intellectual, social and physical well-being for their fellow-men. The main weapon in our own hands is the advancement of our own science (p. 10).

The potential range of such advances is great, especially when compared with the likely narrowness of any one step forwards. Little wonder, then, that geographers find so many different paths to follow.

Human geography has no single disciplinary matrix at the present time, therefore, and has not had one during the period since the Second World War. Rather there have been several competing for a stable position, if not dominance, both within the discipline and beyond. Each matrix has its own branches with their particular exemplars and their leaders who chart progress and seek influence over the whole discipline. At times, the number of branches and their lack of cohesion, especially in the teaching sphere, may suggest anarchy (Johnston, 1976b). In a science about the complexity of people, and organized by complex people, perhaps the existence of such an anarchic situation (or at least the tendency towards it) is all that can be expected. There are schools of thought which wax and wane, some linked to others, some independent; but there is no consensus, no paradigm-dominance, only a series of mutual accommodations which reflects the liberal democratic societal setting of modern Anglo-American human geography.

And the future?

Clearly, given the preceding conclusion, to predict the future of human geography—which schools of thought will contract soon, which will expand, and what new ones will be established—is an impossible task. A combination of external circumstances, internal structures and personal dispositions will determine the discipline's evolution. Prediction involving these factors is hazardous indeed, although it might be reasonable to suggest that the continued numerical dominance of the spatial science disciplinary matrix among the discipline's personnel during the late 1970s will have a major influence on the form of the proximate future. (The continued economic difficulties of western societies, and their reflection in resource allocation to research and higher education, will undoubtedly assist this.)

Numerous geographers are, of course, charting what they would like to see as the contents of human geography in the 1980s. Both Guelke (1977b) and Gregory (1978a), for example, have called for a resuscitation of regional geography, though in markedly different forms. Gregory sees regional geography as providing the understanding of what operates the cam-shafts (Figure 6.1) in particular situations, and incorporates it within a critical science of human geography which is 'committed to emancipation' (p. 70). This will combine the hermeneutic approaches, aimed at improving mutual understanding at the individual level, with structural efforts intended to identify the basic economic and social processes that impinge on the individual: together these will allow for a committed explanation geared to achieving social change through social understanding, and will replace geography's current 'inchoate eclectic-ism' (p. 169) by a structured whole which is a part of an integrated social science. Hay (1979) similarly proposed an integration of the currently competing disciplinary matrices, incorporating not only the hermeneutic and structural approaches but also the positivist spatial science, aban-doned by Gregory: Hay's schema will allow understanding of how the world presently is operating, how it impinges on the individual, and how the whole economic and social system is evolving. (For examples, see also Johnston, 1979.)

Others have their own idiosyncratic views, as reflected in some of the chapter titles in a book *Directions in Geography* (Chorley, 1973) designed to ask those involved in the spatial science developments to review those changes but which became a series of statements about 'the many poss-ible directions which our discipline may follow in the future . . . this volume . . . [indicates] that the recent surge of vitality in our discipline has led to a healthy proliferation of attitudes and objectives' (pp. xi–xii). These views cover not only research activities but also teaching strategies, as exemplified by Gould's (1977, 1978) essays on the topic in the recently launched *Journal of Geography in Higher Education* and elsewhere. The titles in Chorley's book include 'A paradigm for modern geography' (B. J. L. Berry); 'The domain of human geography' (T. Hägerstrand); 'New geography as general spatial systems theory—old social physics writ large' (W. Warntz); 'Some questions about spatial distributions' (M. F. Dacey); 'Geography as human ecology' (R. J. Chorley); 'Future geographies' (W. L. Garrison); and 'Ethics and logic in geography' (W. Bunge). Merely taking these as a representative sample of the views of (at least past) iconoclasts can only lead to the view that human geography will continue 'branching towards anarchy'.

Bibliography

ABLER, R. F. 1971: Distance, intercommunications, and geography. *Proceedings, Association of American Geographers* 3, 1–5.

ABLER, R. F., ADAMS, J. S. and GOULD, P. R. 1971: *Spatial Organization: The Geographer's View of the World*. Englewood Cliffs: Prentice-Hall.

ACKERMAN, E. A. 1945: Geographic training, wartime research, and immediate professional objectives. *Annals, Association of American Geographers* 35, 121–43.

ACKERMAN, E. A. 1958: *Geography as a Fundamental Research Discipline*. University of Chicago, Department of Geography Research Paper 53.

ACKERMAN, E. A. 1963: Where is a research frontier? *Annals, Association of American Geography* 53, 429–40.

ADAMS, J. S. 1969: Directional bias in intra-urban migration. *Economic Geography* 45, 302–23.

ALEXANDER, J. W. and ZAHORCHAK, G. A. 1943: Population-density maps of the United States: techniques and patterns. *Geographical Review* 33, 457–66.

AMEDEO, D. and GOLLEDGE, R. G. 1975: *An Introduction to Scientific Reasoning in Geography*. New York: John Wiley.

ANDERSON, J. 1973: Ideology in geography: an introduction. *Antipode* 5(3), 1–6.

APPLETON, J. 1975: *The Experience of Landscape*. London: John Wiley.

BADCOCK, B. A. 1970: Central-place evolution and network development in south Auckland, 1840–1968: a systems analytic approach. *New Zealand Geographer* 26, 109–35.

BAKER, A. R. H. 1972: Rethinking historical geography. In A. R. H. Baker (editor), *Progress in Historical Geography*, Newton Abbott: David and Charles, 11–28.

BALLABON, M. B. 1957: Putting the 'economic' into economic geography. *Economic Geography* 33, 217–23.

BARROWS, H. H. 1923: Geography as human ecology. *Annals, Association of American Geographers* 13, 1–14.

BATTY, M. 1976: *Urban Modelling: Algorithms, Calibrations, Predictions*. London: Cambridge University Press.

BATTY, M. 1978: Urban models in the planning process. In D. T. Herbert and R. J. Johnston (editors), *Geography and the Urban Environment, Volume 1*, London: Wiley, 63–134.

BELL, D. 1973: *The Coming of Post-Industrial Society*. New York: Basic Books.

BENNETT, R. J. 1974: Process identification for time-series modelling in urban and regional planning. *Regional Studies* 8, 157–74.

BENNETT, R. J. 1975: Dynamic systems modelling of the Northwest region: 1. Spatio-temporal representation and identification. 2. Estimation of the spatio-temporal policy model. 3. Adaptive parameter policy model. 4. Adaptive spatio-temporal forecasts. *Environment and Planning A* 7, 525–38, 539–66, 617–36, 887–98.

BENNETT, R. J. 1978a: Forecasting in urban and regional planning closed loops: the examples of road and air traffic forecasts. *Environment and Planning A* 10, 145–62.

BENNETT, R. J. 1978b: *Spatial Time Series: Analysis, Forecasting and Control*, London: Pion Ltd.

BENNETT, R. J. 1979: Space-time models and urban geographical research in D. T. Herbert and R. J. Johnston (editors), *Geography and the Urban Environment: Progress in Research and Application*. Volume 2. London: John Wiley, 27–58.

BENNETT, R. J. and CHORLEY, R. J. 1978: *Environmental Systems: Philosophy, Analysis and Control*. London: Methuen.

BERRY, B. J. L. 1959a: Ribbon developments in the urban business pattern. *Annals, Association of American Geographers* 49, 145–55.

BERRY, B. J. L. 1959b: Further comments concerning 'geographic' and 'economic' economic geography. *The Professional Geographer* 11(1), 11–12.

BERRY, B. J. L. 1964a: Cities as systems within systems of cities. *Papers, Regional Science Association* 13, 147–63.

BERRY, B. J. L. 1964b: Approaches to regional analysis: a synthesis. *Annals, Association of American Geographers* 54, 2–11.

BERRY, B. J. L. 1965: Research frontiers in urban geography. In P. M. Hauser and L. F. Schnore (editors), *The Study of Urbanization*. New York: Wiley, 403–30.

BERRY, B. J. L. 1966: *Essays on Commodity Flows and the Spatial Structure of the Indian Economy*. University of Chicago, Department of Geography, Research Paper 111.

BERRY, B. J. L. 1967: *The Geography of Market Centers and Retail Distribution*. Englewood Cliffs: Prentice-Hall.

BERRY, B. J. L. 1968: A synthesis of formal and functional regions using a general field theory of spatial behavior. In B. J. L. Berry and D. F. Marble (editors) *Spatial Analysis*, Englewood Cliffs: Prentice-Hall, 419–28.

BERRY, B. J. L. (editor) 1971: *Comparative Factorial Ecology. Economic Geography* 47.

BERRY, B. J. L. 1972a: Hierarchical diffusion: the basis of developmental filtering and spread in a system of growth centers. In N. M. Hansen (editor), *Growth Centers in Regional Economic Development*, New York: The Free Press, 108–38.

BERRY, B. J. L. 1972b: 'Revolutionary and counter revolutionary theory in geography'—a ghetto commentary. *Antipode* 4(2), 31–3.

BERRY, B. J. L. 1972c: More on relevance and policy analysis. *Area* 4, 77–80.

BERRY, B. J. L. 1973a: *The Human Consequences of Urbanization*. London: Macmillan.

BERRY, B. J. L. 1973b: A paradigm for modern geography. In R. J. Chorley (editor), *Directions in Geography*, London: Methuen, 3–22.

BERRY, B. J. L. 1974a: Review of H. M. Rose (editor), *Perspectives in Geography 2. Geography of the Ghetto, Perceptions, Problems and Alternatives. Annals, Association of American Geographers* 64, 342–5.

BERRY, B. J. L. 1974b: Review of David Harvey, *Social Justice and the City. Antipode* 6(2), 142–5, 148.

BERRY, B. J. L. and BAKER, A. M. 1968: Geographic sampling. In B. J. L. Berry and D. F. Marble (editors), *Spatial Analysis*, Englewood Cliffs: Prentice-Hall, 91–100.

BERRY, B. J. L. and GARRISON, W. L. 1958c: The functional bases of the central place hierarchy. *Economic Geography* 34, 145–54.

BERRY, B. J. L. and GARRISON, W. L. 1958b: Recent developments in central place theory. *Papers and Proceedings, Regional Science Association* 4, 107–20.

BERRY, B. J. L. and HORTON, F. E. (editors) 1974: *Urban Environmental Management: Planning for Pollution Control*. Englewood Cliffs: Prentice-Hall.

BERRY, B. J. L. *et al*. 1974: *Land Use, Urban Form and Environmental Quality*. Department of Geography, Research Paper 155, University of Chicago.

VON BERTALANFFY, L. 1950: An outline of general systems theory. *British Journal of the Philosophy of Science* 1, 134–65.

BILLINGE, M. 1977: In search of negativism: phenomenology and historical geography. *Journal of Historical Geography* 3, 55–68.

BIRD, J. H. 1975: Methodological implications for geography from the philosophy of K. R. Popper. *Scottish Geographical Magazine* 91, 153–63.

BIRD, J. H. 1977: Methodology and philosophy. *Progress in Human Geography* 1, 104–10.

BIRD, J. H. 1978: Methodology and philosophy. *Progress in Human Geography* 2, 133–40.

BLOWERS, A. T. 1972: Bleeding hearts and open values. *Area* 4, 290–2.

BLOWERS, A. T. 1974: Relevance, research and the political process. *Area* 6, 32–6.

BODDY, M. J. 1976: The structure of mortgage finance: building societies and the British social formation. *Transactions, Institute of British Geographers* NSI, 58–71.

BRITTAN, S. 1977: Economic liberalism. In A. Bullock and O. Stallybrass (editors), *The Fontana Dictionary of Modern Thought*, London: Fontana Books, 188–9.

BROOKFIELD, H. C. 1962: Local study and comparative method: an example from Central New Guinea. *Annals, Association of American Geographers* 52, 242–54.

BROOKFIELD, H. C. 1964: Questions on the human frontiers of geography. *Economic Geography* 40, 283–303.

BROOKFIELD, H. C. 1969: On the environment as perceived. In C. Board *et al.* (editors), *Progress in Geography* 1, London: Edward Arnold, 51–80.

BROOKFIELD, H. C. 1973: On one geography and a Third World. *Transactions, Institute of British Geographers* 58, 1–20.

BROOKFIELD, H. C. 1975: *Interdependent Development*. London: Methuen.

BROWN, L. A. 1968: *Diffusion Processes and Location: A Conceptual Framework and Bibliography*. Regional Science Research Institute, Bibliography Series three, Philadelphia.

BROWN, L. A. 1975: The market and infrastructure context of adoption: a spatial perspective on the diffusion of innovation. *Economic Geography* 51, 185–216.

BROWN, L. A. and MOORE, E. G. 1970: The intra-urban migration process: a perspective. *Geografiska Annaler* 528, 1–13.

BROWN, S. E. 1978: Guy-Harold Smith, 1895–1976. *Annals, Association of American Geography* 68, 115–18.

BRUNN, S. D. 1974: *Geography and Politics in America*. New York: Harper and Row.

BRUSH, J. E. 1953: The hierarchy of central places in southwestern Wisconsin. *Geographical Review* 43, 380–402.

BUCHANAN, K. 1962: West wind, east wind. *Geography* 47, 333–46.

BUCHANAN, K. 1973: The white north and the population explosion. *Antipode* 5(3), 7–15.

BULLOCK, A. 1977: Liberalism. In A. Bullock and O. Stallybrass (editors), *The Fontana Dictionary of Modern Thought*, London: Fontana Books, 347.

BUNGE, W. 1962: second edition, 1966. *Theoretical Geography*. Lund Studies in Geography, Series C 1, Lund: C. W. K. Gleerup.

BUNGE, W. 1968: Fred K. Schaefer and the science of geography. *Harvard Papers in Theoretical Geography*, Special Papers Series, Paper A, Laboratory for Computer Graphics and Spatial Analysis, Harvard University, Cambridge, Mass.

BUNGE, W. 1971: *Fitzgerald: Geography of a Revolution*. Cambridge, Mass.: Schlenkman.

BUNGE, W. 1973a: Spatial prediction. *Annals, Association of American Geographers* 63, 566-8.

BUNGE, W. 1973b: Ethics and logic in geography. In R. J. Chorley (editor), *Directions in Geography*, London: Methuen, 317-31.

BUNGE, W. 1973c: The geography of human survival. *Annals, Association American Geographers* 63, 275-95.

BUNGE, W. 1973d: The geography. *The Professional Geographer* 25, 331-7.

BUNGE, W. and BORDESSA, R. 1975: *The Canadian Alternative: Survival, Expeditions and Urban Change*. Geographical Monographs, Atkinson College, York University, Downsview, Ontario.

BURTON, I. 1963: The quantitative revolution and theoretical geography. *The Canadian Geographer* 7, 151-62.

BUTTIMER, A. 1971: *Society and Milieu in the French Geographical Tradition*. Chicago: Rand McNally.

BUTTIMER, A. 1974: *Values in Geography*. Commission on College Geography, Resource Paper 24, Association of American Geographers, Washington.

BUTTIMER, A. 1976: Grasping the dynamism of lifeworld. *Annals, Association of American Geographers* 66, 277-92.

BUTTIMER, A. 1978a: Charism and context: the challenge of *La Géographie Humaine*. In D. Ley and M. S. Samuels (editors), *Humanistic Geography: Prospects and Problems*. Chicago: Maaroufa Press, 58-76.

BUTTIMER, A. 1978b: On people, paradigms and progress in geography. Institutionen for Kulturgeografi och Economisk Geografi vid Lunds Universitet, *Rapporter och Notiser* 47.

CAREY, H. C. 1858: *Principles of Social Science*. Philadelphia: J. Lippincott.

CARLSTEIN, T., PARKES, D. N. and THRIFT, N. J. (editors) 1978: *Timing Space and Spacing Time* (three volumes) London: Edward Arnold.

CARROTHERS, G. A. P. 1956: An historical review of the gravity and potential concepts of human interaction. *Journal, American Institute of Planners* 22, 94-102.

CHAPMAN, G. P. 1977: *Human and Environmental Systems: A Geographer's Appraisal*. London: Academic Press.

CHAPPELL, J. E. Jr. 1975: The ecological dimension: Russian and American views. *Annals, Association of American Geographers* 65, 144-62.

CHAPPELL, J. E. Jr. 1976: Comment in reply. *Annals, Association of American Geographers* 66, 169-73.

CHAPPELL, J. M. A. and WEBBER, M. J. 1970: Electrical analogues of spatial diffusion processes. *Regional Studies* 4, 25-39.

CHISHOLM, M. 1962: *Rural Settlement and Land Use*. London: Hutchinson.

CHISHOLM, M. 1966: *Geography and Economics*. London: G. Bell and Sons.

CHISHOLM, M. 1967: General systems theory and geography. *Transactions, Institute of British Geographers* 42, 45–52.

CHISHOLM, M. 1971a: In search of a basis for location theory: microeconomics or welfare economics? In C. Board *et al*. (editors), *Progress in Geography* 3, London: Edward Arnold, 111–34.

CHISHOLM, M. 1971b: Geography and the question of 'relevance'. *Area* 3, 65–8.

CHISHOLM, M. 1973: The corridors of geography. *Area* 5, 43.

CHISHOLM, M. 1975: *Human Geography: Evolution or Revolution?* Harmondsworth: Penguin Books.

CHISHOLM, M. 1976: Regional policies in an era of slow population growth and higher unemployment. *Regional Studies* 10, 201–13.

CHISHOLM, M., FREY, A. E. and HAGGETT, P. (editors), 1971: *Regional Forecasting*. London: Butterworth.

CHISHOLM, M. and MANNERS, G. (editors) 1973: *Spatial Policy Problems of the British Economy*. London: Cambridge University Press.

CHORLEY, R. J. 1962: Geomorphology and general systems theory. *Professional Paper* 500-B, United States Geological Survey, Washington.

CHORLEY, R. J. 1964: Geography and analogue theory. *Annals, Association of American Geographers* 54, 127–37.

CHORLEY, R. J. 1973a: Geography as human ecology. In R. J. Chorley (editor), *Directions in Geography*, London: Methuen, 155–70.

CHORLEY, R. J. (editor) 1973b: *Directions in Geography*. London: Methuen.

CHORLEY, R. J. and HAGGETT, P. 1965a: Trend-surface mapping in geographical research. *Transactions and Papers, Institute of British Geographers* 37, 47–67.

CHORLEY, R. J. and HAGGETT, P. (editors) 1965b: *Frontiers in Geographical Teaching*. London: Methuen.

CHORLEY, R. J. and HAGGETT, P. (editors) 1967: *Models in Geography*. London: Methuen.

CHORLEY, R. J. and KENNEDY, B. A. 1971: *Physical Geography: A Systems Approach*. London: Prentice-Hall International.

CHRISTALLER, W. 1966: *Central Places in Southern Germany* (translated by C. W. Baskin). Englewood Cliffs: Prentice-Hall.

CLARK, A. H. 1954: Historical geography. In P. E. James and C. F. Jones (editors), *American Geography: Inventory and Prospect*, Syracuse: Syracuse University Press, 70–105.

CLARK, A. H. 1977: The whole is greater than the sum of the parts: a humanistic element in human geography. In D. R. Deskins *et al*. (editors), *Geographic Humanism, Analysis and Social Action: A Half Century of Geography of Michigan*, Michigan Geographical Publication No. 17, Ann Arbor, 3–26.

CLARK, D., DAVIES, W. K. D. and JOHNSTON, R. J. 1974: The application of factor analysis in human geography. *The Statistician* 23, 259–81.

CLARK, K. G. T. 1950: Certain underpinnings of our arguments in human geography. *Transactions, Institute of British Geography* 16, 15–22.

CLARK, W. A. V. 1975: Locational stress and residential mobility in a New Zealand context. *New Zealand Geographer* 31, 67–79.

CLARKSON, J. D. 1970: Ecology and spatial analysis. *Annals, Association of American Geographers* 60, 700–16.

CLIFF, A. D. and ORD, J. K. 1973: *Spatial Autocorrelation*. London: Pion Ltd.

CLIFF, A. D. et al. 1975: *Elements of Spatial Structure: A Quantitative Approach*. London: Cambridge University Press.

COATES, B. E., JOHNSTON, R. J. and KNOX, P. L. 1977: *Geography and Inequality*. London: Oxford University Press.

COLE, J. P. 1969: Mathematics and geography. *Geography* 54, 152–63.

COLE, J. P. and KING, C. A. M. 1968: *Quantitative Geography*. London: John Wiley.

COOPER, W. 1952: *The Struggles of Albert Woods*. London: Jonathan Cape.

COPPOCK, J. T. 1974: Geography and public policy: challenges, opportunities and implications. *Transactions, Institute of British Geographers* 63, 1–16.

COURT, A. 1972: All statistical populations are estimated from samples. *The Professional Geographer* 24, 160–1.

COX, K. R. 1969: The voting decision in a spatial context. In C. Board *et al.* (editors), *Progress in Geography* 1, London: Edward Arnold, 81–118.

COX, K. R. 1973: *Conflict, Power and Politics in the City: A Geographic View*. New York: McGraw Hill.

COX, K. R. 1976: American geography: social science emergent. *Social Science Quarterly* 57, 182–207.

COX, K. R. and GOLLEDGE, R. G. 1969: Editorial introduction: behavioral models in geography. In K. R. Cox and R. G. Golledge (editors), *Behavioral Problems in Geography: A Symposium*. Northwestern University Studies in Geography 17, Evanston, 1–13.

COX, K. R., REYNOLDS, D. R. and ROKKAN, S. (editors) 1974: *Locational Approaches to Power and Conflict*. New York: Halsted Press.

CRANE, D. 1972: *Invisible Colleges*. Chicago: University of Chicago Press.

CROWE, P. R. 1936: The rainfall regime of the Western Plains. *Geographical Review* 26, 463–84.

CROWE, P. R. 1938: On progress in geography. *Scottish Geographical Magazine* 54, 1–19.

198   *Geography and Geographers*

CROWE, P. R. 1970: Review of *Progress in Geography* 1. *Geography* 55, 346–7.

CUMBERLAND, K. B. 1947: *Soil Erosion in New Zealand*. Wellington: Whitcombe and Tombs.

CURRY, L. 1967: Quantitative geography. *The Canadian Geographer* 11, 265–74.

CURRY, L. 1972: A spatial analysis of gravity flows. *Regional Studies* 6, 131–47.

DACEY, M. F. 1962: Analysis of central-place and point patterns by a nearest-neighbor method. In K. Norborg (editor), *Proceedings of the IGU Symposium in Urban Geography, Lund 1960*, Lund: C. W. K. Gleerup, 55–76.

DACEY, M. F. 1968: A review on measures of contiguity for two and k-color maps. In B. J. L. Berry and D. F. Marble (editors), *Spatial Analysis*, Englewood Cliffs: Prentice-Hall, 479–95.

DACEY, M. F. 1973: Some questions about spatial distributions. In R. J. Chorley (editor), *Directions in Geography*, London: Methuen, 127–52.

DARBY, H. C. 1953: On the relations of geography and history. *Transactions and Papers, Institute of British Geographers* 19, 1–11.

DARBY, H. C. (editor) 1973: *A New Historical Geography of England*. London: Cambridge University Press.

DARBY, H. C. 1977: *Domesday England*. London: Cambridge University Press.

DAVIES, W. K. D. 1972: Geography and the methods of modern science. In W. K. D. Davies (editor), *The Conceptual Revolution in Geography*, London: University of London Press, 131–9.

DAVIS, W. M. 1906: An inductive study of the content of geography. *Bulletin of the American Geographical Society* 38, 67–84.

DAYSH, G. H. J. (editor) 1949: *Studies in Regional Planning*. London: Philip and Son.

DEAR, M. and CLARK, G. L. 1978: The state and geographic process: a critical review. *Environment and Planning, A* 10, 173–84.

DETWYLER, T. R. and MARCUS, M. G. (editors) 1972: *Urbanization and Environment*. Belmont, California: Duxbury Press.

DICKENSON, J. P. and CLARKE, C. G. 1972: Relevance and the 'newest geography'. *Area* 4, 25–7.

DICKINSON, R. E. 1947: *City, Region and Regionalism*. London: Routledge and Kegan Paul.

DOWNS, R. M. 1970: Geographic space perception: past approaches and future prospects. In C. Board *et al.* (editors), *Progress in Geography* 2. London: Edward Arnold, 65–108.

DOWNS, R. M. and STEA, D. (editors) 1973: *Image and Environment*. London: Edward Arnold.

DOWNS, R. M. and STEA, D. 1977: *Maps in Mind*. New York: Harper and Row.

DRYSDALE, A. and WATTS, M. 1977: Modernization and social protest movements. *Antipode* 9(1), 40–55.

DUNCAN, O. D. 1959: Human ecology and population studies. In P. M. Hauser and O. D. Duncan (editors), *The Study of Population*, Chicago: University of Chicago, 678–716.

DUNCAN, O. D., CUZZORT, R. P. and DUNCAN, B. 1961: *Statistical Geography*. New York: The Free Press.

DUNCAN, O. D. and SCHNORE, L. F. 1959: Cultural, behavioral and ecological perspectives in the study of social organization. *American Journal of Sociology* 65, 132–46.

DUNCAN, S. S. 1974: Cosmetic planning or social engineering? Improvement grants and improvement areas in Huddersfield. *Area* 6, 259–70.

DUNCAN, S. S. 1975: Research directions in social geography: housing opportunities and constraints. *Transactions, Institute of British Geographers* NSI, 10–19.

EILON, S. 1975: Seven faces of research. *Operational Research Quarterly* 26, 359–67.

ELIOT HURST, M. E. 1972: Establishment geography: or how to be irrelevant in three easy lessons. *Antipode* 5(2), 40–59.

ELIOT HURST, M. E. 1973: Transportation and the societal framework. *Economic Geography* 49, 163–80.

ENTRIKIN, N. J. 1976: Contemporary humanism in geography. *Annals, Association of American Geographers* 66, 615–32.

EYLES, J. 1971: Pouring new sentiments into old theories: how else can we look at behavioural patterns? *Area* 3, 242–50.

EYLES, J. 1973: Geography and relevance. *Area* 5, 158–60.

EYLES, J. 1974: Social theory and social geography. In C. Board *et al.* (editors), *Progress in Geography* 6, London: Edward Arnold, 27–88.

EYRE, S. R. 1978: *The Real Wealth of Nations*. London: Edward Arnold.

FOLKE, S. 1972: Why a radical geography must be Marxist. *Antipode* 4(2), 13–18.

FOLKE, S. 1973: First thoughts on the geography of imperialism. *Antipode* 5(3), 16–20.

FOOTE, D. C. and GREER-WOOTTEN, B. 1968: An approach to systems analysis in cultural geography. *The Professional Geographer* 20, 86–90.

FORER, P. C. 1974: Space through time: a case study with New Zealand airlines. In E. L. Cripps (editor), *Space-Time Concepts in Urban and Regional Models*, London: Pion Ltd, 22–45.

FORRESTER, J. W. 1969: *Urban Dynamics*. Cambridge, Mass.: The MIT Press.

FREEMAN, T. W. 1961: *A Hundred Years of Geography*. London: Gerald Duckworth.

FULLER, G. A. 1971: The geography of prophylaxis: an example of intuitive schemes and spatial competition in Latin America. *Antipode* 3(1), 21–30.

GARRISON, W. L. 1953: Remoteness and the passenger utilization of air transportation. *Annals, Association of American Geographers* 43, 169.

GARRISON, W. L. 1956a: Applicability of statistical inference to geographical research. *Geographical Review* 46, 427–9.

GARRISON, W. L. 1956b: Some confusing aspects of common measurements. *The Professional Geographer* 8, 4–5.

GARRISON, W. L. 1959a: Spatial structure of the economy I. *Annals, Association of American Geographers* 49, 238–9.

GARRISON, W. L. 1959b: Spatial structure of the economy II. *Annals, Association of American Geographers* 49, 471–82.

GARRISON, W. L. 1960a: Spatial structure of the economy III. *Annals, Association of American Geographers* 50, 357–73.

GARRISON, W. L. 1960b: Connectivity of the interstate highway system. *Papers and Proceedings, Regional Science Association* 6, 121–37.

GARRISON, W. L. 1962: Simulation models of urban growth and development. In K. Norborg (editor), *I.G.U. Symposium in Urban Geography*, Lund Studies in Geography B 24, Lund; C. W. K. Gleerup, 91–108.

GARRISON, W. L., BERRY, B. J. L., MARBLE, D. F., NYSTUEN, J. D. and MORRILL, R. L. 1959: *Studies of Highway Development and Geographic Change*. Seattle: University of Washington Press.

GARRISON, W. L. and MARBLE, D. F. (editors) 1967a: *Quantitative Geography. Part I: Economic and Cultural Topics*. Northwestern University Studies in Geography, Number 14, Evanston, Illinois.

GARRISON, W. L. and MARBLE, D. F. (editors): *Quantitative Geography. Part I: Economic and Cultural Topics*. Northwestern University Studies in Geography, Number 14, Evanston, Illinois.

GARRISON, W. L. and MARBLE, D. F. (editors) 1967b: *Quantitative Geography. Part II: Physical and Cartographic Topics*. Northwestern University Studies in Geography, Number 14, Evanston, Illinois.

GEARY, R. C. 1954: The contiguity ratio and statistical mapping. *The Incorporated Statistician* 5, 115–41.

GETIS, A. 1963: The determination of the location of retail activities with the use of a map transformation. *Economic Geography* 39, 1–22.

GETIS, A. and BOOTS, B. N. 1978: *Models of Spatial Processes*. London: Cambridge University Press.

GIBSON, E. 1978: Understanding the subjective meaning of places. In D. Ley and M. S. Samuels (editors), *Humanistic Geography: Prospects and Problems*. Chicago: Maaroufa Press, 138–54.

GINSBURG, N. 1972: The mission of a scholarly society. *The Professional Geographer* 24, 1–6.

GINSBURG, N. 1973: From colonialism to national development: geographical perspectives on patterns and policies. *Annals, Association of American Geographers* 63, 1–21.

GLACKEN, C. J. 1956: Changing ideas of the habitable world. In W. L. Thomas (editors), *Man's Role in Changing the Face of the Earth*, Chicago: University of Chicago Press, 70–92.

GLACKEN, C. J. 1967: *Traces on the Rhodian Shore: nature and culture in Western thought from ancient times to the end of the eighteenth century*. Berkeley: University of California Press.

GLACKEN, C. J. 1978: Foreword. In D. Ley and M. S. Samuels (editors), *Humanistic Geography: Prospects and Problems*. Chicago: Maaroufa Press.

GOLLEDGE, R. G. 1969: The geographical relevance of some learning theories. In K. R. Cox and R. G. Golledge (editors), *Behavioral Problems in Geography: A Symposium*, Evanston: Northwestern University Studies in Geography 17, 101–45.

GOLLEDGE, R. G. 1970: Some equilibrium models of consumer behavior. *Economic Geography* 46, 417–24.

GOLLEDGE, R. G. and AMEDEO, D. 1968: On laws in geography. *Annals, Association of American Geography* 58, 760–74.

GOLLEDGE, R. G. and BROWN, L. A. 1967: Search, learning and the market decision process. *Geografiska Annaler* 49B, 116–24.

GOLLEDGE, R. G., BROWN, L. A. and WILLIAMSON, F. 1972: Behavioural approaches in geography: an overview. *The Australian Geographer* 12, 59–79.

GOULD, P. R. 1963: Man against his environment: a game theoretic framework. *Annals, Associations of American Geographers* 53, 290–7.

GOULD, P. R. 1966: *On Mental Maps*. Michigan Inter-University Community of Mathematical Geographers, Discussion Paper 9. Reprinted in R. M. Downs and D. Stea, 1973, *Image and Environment*, London: Edward Arnold, 182–220.

GOULD, P. R. 1969: Methodological developments since the fifties. In C. Board *et al.* (editors) *Progress in Geography* 1, 1–50.

GOULD, P. R. 1970a Is *statistix inferens* the geographical name for a wild goose? *Economic Geography* 46, 439–48.

GOULD, P. R. 1970b: Tanzania 1920–63: the spatial impress of the modernization process. *World Politics* 22, 149–70.

GOULD, P. R. 1972: Pedagogic review. *Annals, Association of American Geographers* 62, 689–700.

GOULD, P. R. 1975: Mathematics in geography: conceptual revolution or new tool? *International Social Science Journal* 27, 303–27.

GOULD, P. R. 1977: What is worth teaching in geography? *Journal of Geography in Higher Education* 1, 20–36.

GOULD, P. R. 1978: Concerning a geographic education. In D. A. Lanegran and R. Palm (editors), *An Invitation to Geography*, New York: McGraw Hill, 202–26.

GOULD, P. R. and WHITE, R. 1974: *Mental Maps*. Harmondsworth: Penguin Books.

GRAY, F. 1975: Non-explanation in urban geography. *Area* 7, 228–35.

GRAY, F. 1976: Selection and allocation in council housing. *Transactions, Institute of British Geographers* NSI, 34–46.

GREER-WOOTTEN, B. 1972: *The Role of General Systems Theory in Geographic Research*. Department of Geography, York University, Discussion Paper No. 3, Toronto.

GREGORY, D. 1976: Rethinking historical geography. *Area* 8, 295–9.

GREGORY, D. 1978a: *Ideology, Science and Human Geography*. London: Hutchinson.

GREGORY, D. 1978b: The discourse of the past: phenomenology, structuralism, and historical geography. *Journal of Historical Geography* 4, 161–73.

GREGORY, S. 1963: *Statistical Methods and the Geographer*, London: Longman.

GREGORY, S. 1976: On geographical myths and statistical fables. *Transactions, Institute of British Geographers* NSI, 385–400.

GRIGG, D. B. 1977: E. G. Ravenstein and the laws of migration. *Journal of Historical Geography* 3, 41–54.

GROSSMAN, L. 1977: Man-environment relationships in anthropology and geography. *Annals, Association of American Geographers* 67, 126–44.

GUELKE, L. 1971: Problems of scientific explanation in geography. *The Canadian Geographer* 15, 38–53.

GUELKE, L. 1974: An idealist alternative in human geography. *Annals, Association of American Geographers* 14, 193–202.

GUELKE, L. 1975: On rethinking historical geography. *Area* 7, 135–8.

GUELKE, L. 1976: The philosophy of idealism. *Annals, Association of American Geography* 66, 168–9.

GUELKE, L. 1977a: The role of laws in human geography. *Progress in Human Geography* 1, 376–86.

GUELKE, L. 1977b: Regional geography. *The Professional Geographer* 29, 1–7.

GUELKE, L. 1978: Geography and logical positivism. In D. T. Herbert and R. J. Johnston (editors), *Geography and the Urban Environment: Progress in Research and Applications, Volume 1*, London: John Wiley, 35–61.

HÄGERSTRAND, T. 1968: *Innovation Diffusion as a Spatial Process*. Chicago: University of Chicago.

HÄGERSTRAND, T. 1977: The geographers' contribution to regional policy: the case of Sweden. In D. R. Deskins *et al.* (editors) *Geographic*

Humanism, Analysis and Social Action: A Half Century of Geography at Michigan. Michigan Geographical Publications No. 17, Ann Arbor, 329–46.

HAGGETT, P. 1964: Regional and local components in the distribution of forested areas in southeast Brazil: a multivariate approach. *Geographical Journal* 130, 365–77.

HAGGETT, P. 1965a: Changing concepts in economic geography. In R. J. Chorley and P. Haggett (editors), *Frontiers in Geographical Teaching*, London: Methuen, 101–17.

HAGGETT, P. 1965b: Scale components in geographical problems. In R. J. Chorley and P. Haggett (editors), *Frontiers in Geographical Teaching*, London: Methuen. 164–85.

HAGGETT, P. 1965c: *Locational Analysis in Human Geography*. London: Edward Arnold.

HAGGETT, P. 1973: Forecasting alternative spatial, ecological and regional futures: problems and possibilities. In R. J. Chorley (editor), *Directions in Geography*, London: Methuen, 217–36.

HAGGETT, P. and CHORLEY, R. J. 1965: Frontier movements and the geographic tradition. In R. J. Chorley and P. Haggett (editors), *Frontiers in Geographical Teaching*, London: Methuen, 358–78.

HAGGETT, P. and CHORLEY, R. J. 1967: Models, paradigms, and the new geography. In R. J. Chorley and P. Haggett (editors), *Models in Geography*, London: Methuen, 19–42.

HAGGETT, P. and CHORLEY, R. J. 1969: *Network Models in Geography*. London: Edward Arnold.

HAGGETT, P., CLIFF, A. D. and FREY, A. 1977: *Locational Analysis in Human Geography*. London: Edward Arnold.

HAGOOD, M. J. 1943: Development of a 1940 rural farm level of living index for counties. *Rural Sociology* 8, 171–80.

HALL, P. 1974: The new political geography. *Transactions, Institute of British Geographers* 63, 48–52.

HALVORSON, P. and STAVE, B. M. 1978: A conversation with Brian J. L. Berry, *Journal of Urban History* 4, 209–38.

HAMILTON, F. E. I. 1974: A view of spatial behaviour, industrial organizations, and decision-making. In F. E. I. Hamilton (editor), *Spatial Perspectives on Industrial Organization and Decision-Making*, London: John Wiley, 3–46.

HAMNETT, C. 1977: Non-explanation in urban geography: throwing the baby out with the bath water. *Area* 9, 143–5.

HARE, F. K. 1974: Geography and public policy: a Canadian view. *Transactions, Institute of British Geographers* 63, 25–8.

HARE, F. K. 1977: Man's world and geographers: a secular sermon. In D. R. Deskins *et al.* (editors), *Geographic Humanism, Analysis and Social Action: A Half Century of Geography at Michigan*, Michigan Geographical Publication No. 17, Ann Arbor, 259–73.

HARRIES, K. D. 1974: *The Geography of Crime and Justice*. New York: McGraw Hill.

HARRIES, K. D. 1975: Rejoinder to Richard Peet: 'The geography of crime: a political critique.' *The Professional Geographer* 27, 280–2.

HARRIES, K. D. 1976: Observations on radical versus liberal theories of crime causation. *The Professional Geographer* 28, 100–13.

HARRIS, R. C. 1971: Theory and synthesis in historical geography. *The Canadian Geographer* 15, 157–72.

HARRIS, R. C. 1977: The simplification of Europe overseas. *Annals, Association of American Geographers* 67, 469–83.

HARRIS, R. C. 1978: The historical mind and the practice of geography. In D. Ley and M. S. Samuels (editors), *Humanistic Geography: Prospects and Problems*. Chicago: Maaroufa Press, 123–37.

HARRIS, C. D. 1954a: The geography of manufacturing. In P. E. James and C. F. Jones (editors), *American Geography: Inventory and Prospect*, Syracuse: Syracuse University Press, 292–309.

HARRIS, C. D. 1954b: The market as a factor in the localization of industry in the United States. *Annals, Association of American Geographers* 44, 315–48.

HARRIS, C. D. 1977: Edward Louis Ullman, 1912–1976. *Annals, Association of American Geographers* 67, 595–600.

HARRIS, C. D. and ULLMAN, E. L. 1945: The nature of cities. *Annals of the American Academy of Political and Social Science* 242, 7–17.

HARTSHORNE, R. 1939: *The Nature of Geography*. Lancaster, Pennsylvania: Association of American Geographers.

HARTSHORNE, R. 1948: On the mores of methodological discussion in American geography. *Annals, Association of American Geographers* 38, 492–504.

HARTSHORNE, R. 1954a: Political geography. In P. E. James and C. F. Jones (editors), *American Geography: Inventory and Prospect*, Syracuse: Syracuse University Press, 167–225.

HARTSHORNE, R. 1954b: Comment on 'Exceptionalism in geography'. *Annals, Association of American Geographers* 44, 108–9.

HARTSHORNE, R. 1955: 'Exceptionalism in geography' re-examined. *Annals, Association of American Geographers* 45, 205–44.

HARTSHORNE, R. 1958: The concept of geography as a science of space, from Kant and Humboldt to Hettner. *Annals, Association of American Geographers* 48, 97–108.

HARTSHORNE, R. 1954: *Perspective on the Nature of Geography*. Chicago: Rand McNally.

HARTSHORNE, R. 1972: Review of 'Kant's concept of geography'. *The Canadian Geographer* 16, 77–9.

HARVEY, D. 1967a: Models of the evolution of spatial patterns in geography. In R. J. Chorley and P. Haggett (editors), *Models in Geography*, London: Methuen, 549–608.

HARVEY, D. 1967b: Editorial introduction: the problem of theory construction in geography. *Journal of Regional Science* 7, 211–16.
HARVEY, D. W. 1969a: *Explanation in Geography*. London: Edward Arnold.
HARVEY, D. 1969b: Review of A. Pred, *Behavior and Location Part I*. *Geographical Review* 59, 312–14.
HARVEY, D. 1969c: Conceptual and measurement problems in the cognitive-behavioral approach to location theory. In K. R. Cox and R. G. Colledge (editors), *Behavioral Problems in Geography: A Symposium*, Northwestern University Studies in Geography 17, 35–68.
HARVEY, D. 1970: Behavioural postulates and the construction of theory in human geography. *Geographica Polonica* 18, 27–46.
HARVEY, D. 1972: Revolutionary and counter-revolutionary theory in geography and the problem of ghetto formation. *Antipode* 4(2), 1–13.
HARVEY, D. 1973: *Social Justice and the City*. London: Edward Arnold.
HARVEY, D. 1974a: A commentary on the comments. *Antipode* 4(2), 36–41.
HARVEY, D. 1974b: Discussion with Brian Berry. *Antipode* 6(2), 145–8.
HARVEY, D. 1974c: What kind of geography for what kind of public policy? *Transactions, Institute of British Geographers* 63, 18–24.
HARVEY, D. 1974d: Population, resources and the ideology of science. *Economic Geography* 50, 256–77.
HARVEY, D. 1974e: Class-monopoly rent, finance capital and the urban revolution. *Regional Studies* 8, 239–55.
HARVEY, D. 1975a: Class structure in a capitalist society and the theory of residential differentiation. In R. Peel, M. Chisholm and P. Haggett (editors), *Processes in Physical and Human Geography: Bristol Essays*. London: Heinemann, 354–69.
HARVEY, D. 1975b: The political economy of urbanization in advanced capitalist societies: the case of the United States. In G. Gappert and H. M. Rose (editors), *The Social Economy of Cities*, Beverly Hills: Sage Publications, 119–63.
HARVEY, D. 1975c: Review of B. J. L. Berry, *The Human Consequences of Urbanisation*. *Annals, Association of American Geographers* 65, 99–103.
HARVEY, D. 1976: The Marxist theory of the state. *Antipode* 8(2), 80–9.
HARVEY, D. 1978: The urban process under capitalism: a framework for analysis. *International Journal of Urban and Regional Research* 2, 101–32.
HAY, A. M. 1978: Some problems in regional forecasting. In J. I. Clarke and J. Pelletser (editors), *Régions Géographiques et Régions d'Aménagement*. Collection les hommes et les lettres, 7, Editions Hermes, Lyon.
HAY, A. M. 1979: Positivism in human geography: response to critics. In D. T. Herbert and R. J. Johnston (editors), *Geography and the Urban Environment: Progress in Research and Applications*, Volume 2, London: John Wiley, 1–26.

206   *Geography and Geographers*

HAYNES, R. M. 1975: Dimensional analysis: some applications in human geography. *Geographical Analysis* 7, 51–68.

HERBERT, D. T. and JOHNSTON, R. J. 1978: Geography and the urban environment. In D. T. Herbert and R. J. Johnston (editors), *Geography and the Urban Environment: Progress in Research and Applications, Volume 1*, London: John Wiley, 1–29.

HERBERTSON, A. J. 1905: The major natural regions, *Geographical Journal* 25, 300–10.

HODGART, R. L. 1978: Optimizing access to public services: a review of problems, models, and methods of locating central facilities. *Progress in Human Geography* 2, 17–48.

HOOK, J. C. 1955: Areal differentiation of the density of the rural farm population in the northeastern United States. *Annals, Association of American Geographers* 45, 189–90.

HOUSE, J. W. 1973: Geographers, decision takers and policy makers. In M. Chisholm and B. Rodgers (editors), *Studies in Human Geography*, London: Heinemann, 272–305.

ISARD, W. 1956a: *Location and Space Economy*. New York: John Wiley.

ISARD, W. 1956b: Regional science, the concept of region, and regional structure. *Papers and Proceedings, Regional Science Association* 2, 13–39.

ISARD, W. 1960: *Methods of Regional Analysis: An Introduction in Regional Science*. New York: John Wiley.

ISARD, W. et al. 1969: *General Theory: Social, Political Economic and Regional*. Cambridge, Mass.: The MIT Press.

ISARD, W. 1975: *An Introduction to Regional Science*. Englewood Cliffs: Prentice-Hall.

JAMES, P. E. 1942: *Latin America*. London: Cassell.

JAMES, P. E. 1954: Introduction: the field of geography. In P. E. James and C. F. Jones (editors), *American Geography: Inventory and Prospect*, Syracuse: Syracuse University Press, 2–18.

JAMES, P. E. 1965: The President's session. *The Professional Geographer* 17(4), 35–7.

JAMES, P. E. 1972: *All Possible Worlds: A History of Geographical Ideas*. Indianapolis: The Odyssey Press.

JAMES, P. E. and JONES, C. F. (editors) 1954: *American Geography: Inventory and Prospect*. Syracuse: Syracuse University Press.

JANELLE, D. G. 1968: Central-place development in a time-space framework. *The Professional Geographer* 20, 5–10.

JANELLE, D. G. 1969: Spatial reorganization: a model and concept. *Annals, Association of American Geographers* 59, 348–64.

JOHNSTON, R. J. 1969: Urban geography in New Zealand 1945–1969. *New Zealand Geographer* 25, 121–35.

JOHNSTON, R. J. 1971: *Urban Residential Patterns: An Introductory Review*. London: G. Bell and Sons.

JOHNSTON, R. J. 1974: Continually changing human geography revisited: David Harvey: *Social Justice and the City*. *New Zealand Geographer* 30, 180–92.

JOHNSTON, R. J. 1976a: *The World Trade System: Some Enquiries into its Spatial Structure*. London: G. Bell and Sons.

JOHNSTON, R. J. 1976b: Anarchy, conspiracy and apathy: the three 'conditions' of geography. *Area* 8, 1–3.

JOHNSTON, R. J. 1978a: *Multivariate Statistical Analysis in Geography: A Primer on the General Linear Model*. London: Longman.

JOHNSTON, R. J. 1978b: *Political, Electoral and Spatial Systems*. London: Oxford University Press.

JOHNSTON, R. J. 1978c: Paradigms and revolutions or evolution: observations on human geography since the Second World War. *Progress in Human Geography* 2, 189–206.

JOHNSTON, R. J. 1979: Urban geography: city structures. *Progress in Human Geography* 3, 133–8.

JOHNSTON, R. J. and HERBERT, D. T. 1978: Introduction. In D. T. Herbert and R. J. Johnston (editors), *Social Areas in Cities: Processes, Patterns and Problems*, London: John Wiley, 1–33.

JONES, E. 1956: Cause and effect in human geography. *Annals, Association of American Geographers*, 46, 369–77.

KANSKY, K. J. 1963: *Structure of Transport Networks*. Univeristy of Chicago, Department of Geography, Research Paper 84, Chicago.

KASPERSON, R. E. 1971: The post-behavioral revolution geography. *British Columbia Geographical Series* 12, 5–20.

KATES, R. W. 1962: *Hazard and Choice Perception in Flood Plain Management*. University of Chicago, Department of Geography, Research Paper 78, Chicago.

KATES, R. W. 1972: Review of *Perspectives on Resource Management*. *Annals, Association of American Geographers* 62, 519–20.

KING, L. J. 1960: A note on theory and reality. *The Professional Geographer* 12(3), 4–6.

KING, L. J. 1961: A multivariate analysis of the spacing of urban settlement in the United States. *Annals, Association of American Geographers* 51, 222–33.

KING, L. J. 1969: The analysis of spatial form and its relationship to geographic theory. *Annals, Association of American Geographers* 59, 573–95.

KING, L. J. 1976: Alternatives to a positive economic geography. *Annals, Association of American Geographers* 66, 293–308.

KING, L. J. and CLARK, G. L. 1978: Government policy and regional development. *Progress in Human Geography* 2, 1–16.

KIRK, W. 1951: Historical geography and the concept of the behavioural environment. *Indian Geographical Journal* 25, 152–60.

KIRK, W. 1963: Problems of geography. *Geography* 48, 357–71.

KNOS, D. S. 1968: The distribution of land values in Topeka, Kansas. In B. J. L. Berry and D. F. Marble (editors), *Spatial Analysis*, Englewood Cliffs: Prentice-Hall, 269–89.

KNOX, P. L. 1975: *Social Well-Being: A Spatial Perspective*. London: Oxford University Press.

KUHN, T. S. 1962: *The Structure of Scientific Revolutons*. Chicago: University of Chicago Press.

KUHN, T. S. 1970: *The Structure of Scientific Revolutions* (Second Edition). Chicago: University of Chicago Press.

KUHN, T. S. 1977: Second thoughts on paradigms. In F. Suppe (editor), *The Structure of Scientific Theories*, Urbana: University of Illinois Press, 459–82, plus discussion 500–17.

LANGTON, J. 1972: Potentialities and problems of adapting a systems approach to the study of change in human geography. *Progress in Geography* 4, 125–79.

LAVALLE, P., McCONNELL, H. and BROWN, R. G. 1967: Certain aspects of the expansion of quantitative methodology in American geography. *Annals, Association of American Geographers* 57, 423–36.

LAW, J. 1976: Theories and methods in the sociology of science: an interpretative approach. In G. Lemaine *et al*. *Perspectives on the Emergence of Scientific Disciplines*. The Hague: Mouton, 221–31.

LEACH, B. 1974: Race, problems and geography. *Transactions, Institute of British Geographers* 63, 41–7.

LEACH, E. R. 1974: *Lévi-Strauss*. London: Fontana.

LEE, Y. 1975: A rejoinder to 'The geography of crime: a political critique'. *The Professional Geographer* 27, 284–5.

LEMAINE, G. *et al*. 1976: Introduction: problems in the emergence of new disciplines. In G. Lemaine *et al*. (editors), *Perspectives on the Emergence of Scientific Disciplines*, The Hague: Mouton, 1–73.

LEWIS, G. M. 1966: Regional ideas and reality in the Cis-Rocky Mountain West. *Transactions, Institute of British Geographers* 38, 135–50.

LEWIS, G. M. 1968: Levels of living in the Northeastern United States *c*. 1960: a new approach to regional geography. *Transactions, Institute of British Geographers* 45, 11–37.

LEWIS, P. W. 1965: Three related problems in the formulation of laws in geography. *The Professional Geographer* 17(5), 24–7.

LEWTHWAITE, G. R. 1966: Environmentalism and determinism: a search for clarification. *Annals, Association of American Geographers* 56, 1–23.

LEY, D. 1977: The personality of a geographical fact. *The Professional Geographer* 29, 8–13.

LEY, D. 1978: Social geography and social action. In D. Ley and M. S. Samuels (editors), *Humanistic Geography: Prospects and Problems*. Chicago: Maaroufa Press, 41–57.

LEY, D. and SAMUELS, M. S. 1978: Introduction: contexts of modern humanism in geography. In D. Ley and M. S. Samuels (editors), *Humanistic Geography: Prospects and Problems*. Chicago: Maaroufa Press, 1–18.

LÖSCH, A. 1954: *The Economics of Location*. New Haven, Conn.: Yale University Press.

LOWENTHAL, D. 1961: Geography, experience, and imagination: towards a geographical epistemology. *Annals, Association of American Geographers* 51, 241–60.

LOWENTHAL, D. (editor) 1965: *George Perkins Marsh: Man and Nature*. Cambridge, Mass.: Harvard University Press.

LOWENTHAL, D. 1968: The American scene. *Geographical Review* 48, 61–88.

LOWENTHAL, D. and BOWDEN, M. J. (editors) 1975: *Geographies of the Mind: Essays in Historical Geosophy in honor of John Kirkland Wright*. New York: Oxford University Press.

LOWENTHAL, D. and PRINCE, H. C. 1965: English landscape tastes. *Geographical Review* 55, 186–222.

LOWENTHAL, D. *et al.* 1973: Report of the AAG Task Force on environmental quality. *The Professional Geographer* 25, 39–46.

LUKERMANN, F. 1958: Towards a more geographic economic geography. *The Professional Geographer* 10(1), 2–10.

LUKERMANN, F. 1960a: On explanation, model, and prediction. *The Professional Geographer* 12(1), 1–2.

LUKERMANN, F. 1960b: The geography of cement? *The Professional Geographer* 12(4), 1–6.

LUKERMANN, F. 1961: The role of theory in geographical inquiry. *The Professional Geographer* 13(2), 1–6.

LUKERMANN, F. 1965: Geography: de facto or de jure. *Journal of the Minnesota Academy of Science* 32, 189–96.

LYNCH, K. 1960: *The Image of the City*. Cambridge, Mass.: The MIT Press.

McCARTY, H. H. 1940: *The Geographic Basis of American Economic Life*. New York: Harper and Brothers.

McCARTY, H. H. 1952: McCarty on McCarthy: The Spatial Distribution of the McCarthy Vote 1952. Unpublished Paper, Department of Geography, State University of Iowa, Iowa City.

McCARTY, H. H. 1953: An approach to a theory of economic geography. *Annals, Association of American Geographers* 43, 183–4.

McCARTY, H. H. 1954: An approach to a theory of economic geography. *Economic Geography* 30, 95–101.

McCARTY, H. H. 1958: Science, measurement, and area analysis. *Economic Geography* 34, facing page 283.

McCARTY, H. H., HOOK, J. C. and KNOS, D. S. 1956: *The Measurement of Association in Industrial Geography*, Department of Geography, State University of Iowa, Iowa City.

McCARTY, H. H. and LINDBERG, J. B. 1966: *A Preface to Economic Geography*. Englewood Cliffs: Prentice-Hall.

McDANIEL, R. and ELIOT HURST, M. E. 1968: *A Systems Analytic Approach to Economic Geography*. Commission on College Geography, Publication 8, Association of American Geographers, Washington DC.

MACKAY, J. R. 1958: The interactance hypothesis and boundaries in Canada: a preliminary study. *The Canadian Geographer* 11, 1–8.

McKINNEY, W. M. 1968: Carey, Spencer , and modern geography. *The Professional Geographer* 20, 103–6.

McTAGGART, W. D. 1974: Structuralism and universalism in geography: reflections on contributions by H. C. Brookfield. *The Australian Geographer* 12, 510–16.

MABOGUNJE, A. K. 1977: In search of spatial order: geography and the new programme of urbanization in Nigeria. In D. R. Deskins *et al*. (editors) *Geographic Humanism, Analysis and Social Action: A Half Century of Geography at Michigan*. Michigan Geography Publications No. 17, Ann Arbor, 347–76.

MANNERS, I. R. and MIKESELL, M. W. (editors) 1974: *Perspectives on Environment*. Commission on College Geography, Association of American Geographers, Washington.

MARCHAND, B. 1978: A dialectical approach in geography. *Geographical Analysis* 10, 105–19.

MARTIN, A. F. 1951: The necessity for determinism. *Transactions, Institute of British Geographers* 17, 1–12.

MARTIN, R. L. and OEPPEN, J. 1975: The identification of regional forecasting models using space-time correlation functions. *Transactions, Institute of British Geographers* 66, 95–118.

MASSAM, B. H. 1975: *Location and Space in Social Administration*. London: Edward Arnold.

MASTERMAN, M. 1970: The nature of a paradigm. In I. Lakatos and A. Musgrave (editors), *Criticism and the Growth of Knowledge*, London: Cambridge University Press, 59–90.

MAY, J. A. 1970: *Kant's Concept of Geography: and its relation to recent geographical thought*. Department of Geography, University of Toronto, Research Publication 4, Toronto.

MAY, J. A. 1972: A reply to Professor Hartshorne. *The Canadian Geographer*, 16, 79–81.

MAYER, H. M. 1954: Urban geography. In P. E. James and C. F. Jones (editors), *American Geography: Inventory and Prospect*, Syracuse: Syracuse University Press, 142–66.

MEADOWS, D. H. *et al*. 1972: *The Limits of Growth*. New York: Universal Books.

MEINIG, D. W. 1972: American wests: preface to a geographical introduction. *Annals, Association of American Geographers* 62, 159–84.

MERCER, D. C. 1977: *Conflict and Consensus in Human Geography.* Monash Publications in Geography Number 17, Clayton, Victoria, Australia.

MERCER, D. C. and POWELL, J. M. 1972: *Phenomenology and Related Non-Positivistic Viewpoints in the Social Sciences.* Monash Publications in Geography, No. 1, Clayton, Victoria, Australia.

MEYER, D. R. 1972: Geographical population data: statistical description not statistical inference. *The Professional Geographer* 24, 26–8.

MIKESELL, M. W. 1967: Geographical perspectives in anthropology. *Annals, Association of American Geography* 57, 617–34.

MIKESELL, M. W. 1969: The borderlands of geography as a social science. In M. Sherif and C. W. Sherif (editors), *Interdisciplinary Relationships in the Social Sciences,* Chicago: Aldine Publishing Company, 227–48.

MIKESELL, M. W. (editor) 1973: *Geographers Abroad: Essays on the Prospects of Research in Foreign Areas.* Department of Geography, University of Chicago, Research Paper 152, Chicago.

MIKESELL, M. W. 1974: Geography as the study of environment: an assessment of some old and new commitments. In I. R. Manners and M. W. Mikesell (editors), *Perspectives on Environment,* Commission on College Geography, Association of American Geographers, Washington, 1–23.

MIKESELL, M. W. 1978: Tradition and innovation in cultural geography. *Annals, Association of American Geographers* 68, 1–16.

MONTEFIORE, A. G. and WILLIAMS, W. M. 1955: Determinism and possibilism. *Geographical Studies* 2, 1–11.

MOODIE, D. W. and LEHR, J. C. 1976: Fact and theory in historical geography. *The Professional Geographer* 28, 132–6.

MORGAN, W. B. and MOSS, R. P. 1965: Geography and ecology: the concept of the community and its relationship to environment. *Annals, Association of American Geographers* 55, 339–50.

MORRILL, R. L. 1965: *Migration and the Growth of Urban Settlement.* Lund Studies in Geography, Series B, 24, Lund: C. W. K. Gleerup.

MORRILL, R. L. 1968: Waves of spatial diffusion. *Journal of Regional Science* 8, 1–18.

MORRILL, R. L. 1969: Geography and the transformation of society. *Antipode* 1(1), 6–9.

MORRILL, R. L. 1970a: *The Spatial Organization of Society.* Belmont, California: Wadsworth, 2nd edn. 1974.

MORRILL, R. L. 1970b: Geography and the transformation of society: part II. *Antipode* 2(1), 4–10.

MORRILL, R. L. 1974: Review of D. Harvey, *Social Justice and the City. Annals, Association of American Geographers* 64, 475–7.

MORRILL, R. L. and GARRISON, W. L. 1960: Projections of interregional patterns of trade in wheat and flour. *Economic Geography* 36, 116–26.

MORRILL, R. L. and WOHLENBERG, E. H. 1971: *The Geography of Poverty in the United States*. New York: McGraw Hill.

MOSS, R. P. 1970: Authority and charisma: criteria of validity in geographical method. *South African Geographical Journal* 52, 13–37.

MOSS, R. P. 1977: Deductive strategies in geographical generalization. *Progress in Physical Geography* 1, 23–39.

MOSS, R. P. and MORGAN, W. B. 1967: The concept of the community: some applications in geographical research. *Transactions, Institute of British Geographers* 41, 21–32.

MUIR, R. 1975: *Modern Political Geography*. London: Macmillan.

MULKAY, M. J. 1975: Three models of scientific development. *Sociological Review* 23, 509–26.

MULKAY, M. J. 1976: Methodology in the sociology of science: some reflections on the study of radio astronomy. In G. Lemaine *et al.* (editors), *Perspectives in the Emergence of Scientific Disciplines*, The Hague: Mouton, 207–20.

MULKAY, M. J. 1978: Consensus in science. *Social Science Information* 17, 107–22.

MULKAY, M. J., GILBERT, G. N. and WOOLGAR, S. 1975: Problem areas and research networks in science. *Sociology* 9, 187–203.

MURDIE, R. A. 1969: *Factorial Ecology of Metropolitan Toronto 1951–1961*. University of Chicago, Department of Geography, Research Paper 116, Chicago.

MYRDAL, G. 1957: *Economic Theory and Underdeveloped Regions*. London: Duckworth.

NATIONAL ACADEMY OF SCIENCES—NATIONAL RESEARCH COUNCIL 1965: *The Science of Geography*. Washington: NAS—NRC.

NEFT, D. 1966: *Statistical Analysis for Areal Distributions*. Monograph 2, Regional Science Research Institute, Philadelphia.

NEWMAN, J. L. 1973: The use of the term 'hypothesis' in geography. *Annals, Association of American Geographers* 63, 22–7.

NYSTUEN, J. D. 1963: Identification of some fundamental spatial concepts. *Papers of the Michigan Academy of Science, Arts and Letters*, 48, 373–84. Reprinted in B. J. L. Berry and D. F. Marble (editors), *Spatial Analysis*, Englewood Cliffs: Prentice-Hall, 35–41.

ODUM, H. W. and MOORE, H. E. 1938: *American Regionalism—A Cultural-Historical Approach to National Integration*. New York: H. Holt and Company.

OLSSON, G. 1965: *Distance and Human Interaction: A Review and Bibliography*. Regional Science Research Institute, Bibliography Series Number Two, Philadelphia.

OLSSON, G. 1969: Inference problems in locational analysis. In K. R. Cox and R. G. Golledge (editors), *Behavioral Problems in Geography: A Symposium*, Northwestern University Studies in Geography 17, Evanston 14–34.

O'RIORDAN, T. 1971a: Environmental management. In C. Board *et al.* (editors), *Progress in Geography* 3, London: Edward Arnold, 173–231.

O'RIORDAN, T. 1971b: *Perspectives in Resource Management*, London: Pion Ltd.

O'RIORDAN, T. 1976: *Environmentalism*. London: Pion Ltd.

PAHL, R. E. 1965: Trends in social geography. In R. J. Chorley and P. Haggett (editors), *Frontiers in Geographical Teaching*, London: Methuen, 81–100.

PAHL, R. E. 1969: Urban social theory and research. *Environment and Planning* 1, 143–54. Reprinted in R. E. Pahl 1970, *Whose City?* London: Longman, 209–25.

PAHL, R. E. 1975: *Whose City? and Other Essays* Harmondsworth: Penguin Books, (second edition).

PALM, R. 1979: Financial and real estate institutions in the housing market. In D. T. Herbert and R. J. Johnston (editors), *Geography and the Urban Environment*, Volume 2, London: Wiley, 83–124.

PAPAGEORGIOU, G. J. 1969: Description of a basis necessary to the analysis of spatial systems. *Geographical Analysis* 1, 213–15.

PARSONS, J. J. 1977: Geography as exploration and discovery. *Annals, Association of American Geographers* 67, 1–16.

PATERSON, J. H. 1974: Writing regional geography. In C. Board *et al.* (editors), *Progress in Geography* 6, London: Edward Arnold, 1–26.

PATMORE, J. A. 1970: *Land and Leisure*. Newton Abbott: David and Charles.

PEET, J. R. 1971: Poor, hungry America. *The Professional Geographer* 23, 99–104.

PEET, J. R. 1975a: Inequality and poverty: a Marxist-geographic theory. *Annals, Association of American Geographers* 65, 564–71.

PEET, J. R. 1975b: The geography of crime: a political critique. *The Professional Geographer* 27, 277–80.

PEET, J. R. 1976a: Further comments on the geography of crime. *The Professional Geographer* 28, 96–100.

PEET, J. R. 1976b: Editorial: radical geography in 1976. *Antipode* 8(3), inside cover.

PEET, J. R. 1977: The development of radical geography in the United States. *Progress in Human Geography* 1, 240–63.

PEET, J. R. 1978: *Radical Geography*. London: Methuen.

PELTIER, L. C. 1954: Geomorphology. In P. E. James and C. F. Jones (editors), *American Geography: Inventory and Prospect*, Syracuse: Syracuse University Press, 362–81.

PERRY, P. J. 1969: H. C. Darby and historical geography: a survey and review. *Geographische Zeitschrift* 57, 161–77.

PETER, L. and HULL, R. 1969: *The Peter Principle*. London: Bantam Books.

PIRIE, G. H. 1976: Thoughts on revealed preferences and spatial behaviour. *Environment and Planning* A 8, 947–55.

PITTS, F. R. 1965: A graph theoretic approach to historical geography. *The Professional Geographer* 17(5), 15–20.

POCOCK, D. and HUDSON, R. 1978: *Images of the Urban Environment*. London: Macmillan.

POOLER, J. A. 1977: The origins of the spatial tradition in geography: an interpretation. *Ontario Geography* 11, 56–83.

POPPER, K. R. 1959: *The Logic of Scientific Discovery*. London: Hutchinson.

POPPER, K. R. 1967: Replies to my critics. In P. A. Schipp (editor), *The Philosophy of Karl Popper*, Volume 2, La Salle, Indiana: Open Court Publishing Company, 961–97.

POPPER, K. R. 1970: Normal science and its dangers. In I. Lakatos and A. Musgrave (editors), *Criticism and the Growth of Knowledge*, London: Cambridge University Press, 51–8.

PORTER, P. W. and LUKERMANN, F. 1975: The geography of utopia. In D. Lowenthal and M. J. Bowden (editors), *Geographies of the Mind: Essays in Historical Geosophy*, New York: Oxford University Press, 197–224.

POWELL, J. M. 1970: *The Public Lands of Australia Felix: Settlement and Land Appraisal in Victoria 1834–1891*. Melbourne: Oxford University Press.

POWELL, J. M. 1971: Utopia, millenium and the cooperative ideal: a behavioural matrix in the settlement process. *The Australian Geographer* 11, 606–18.

POWELL, J. M. 1972: *Images of Australia*. Monash University Publications in Geography No. 3, Clayton, Victoria, Australia.

POWELL, J. M. 1977: *Mirrors of the New World: Images and Image-Makers in the Settlement Process*. Folkestone: Dawson.

PRED, A. 1965a: The concentration of high value-added manufacturing. *Economic Geography* 41, 108–32.

PRED, A. R. 1965b: Industrialization, initial advantage, and American metropolitan growth. *Geographical Review* 55, 158–85.

PRED, A. 1967: *Behavior and Location: Foundations for a Geographic and Dynamic Location Theory. Part I*. Lund: C. W. K. Gleerup.

PRED, A. 1969: *Behavior and Location: Foundations for a Geographic and Dynamic Location Theory. Part II*. Lund: C. W. K. Gleerup.

PRED, A. 1973: Urbanization, domestic planning problems and Swedish geographic research. In C. Board *et al*. (editors) *Progress in Geography* 5, London: Edward Arnold, 1–77.

PRED, A. 1977: The choreography of existence: comments on Hägerstrand's time-geography and its usefulness. *Economic Geography* 53, 207–21.

PRED, A. and PALM, R. 1978: The status of American women: a time-geographic view. In D. A. Lanegran and R. Palm (editors), *An Invitation to Geography* (second edition), New York: McGraw Hill, 99–109.

PRINCE, H. C. 1971a: Real, imagined and abstract worlds of the past. In C. Board *et al.* (editors), *Progress in Geography* 3, London: Edward Arnold, 1–86.

PRINCE, H. C. 1971b: America! America? Views on a pot melting 1. Questions of social relevance. *Area* 3, 150–3.

RAVENSTEIN, E. G. 1885: The laws of migration. *Journal of the Royal Statistical Society* 48, 167–235.

RAY, D. M., VILLENEUVE, P. Y. and ROBERGE, R. A. 1974: Functional prerequisites, spatial diffusion, and allometric growth. *Economic Geography* 50, 341–51.

REES, P. H. and WILSON, A. G. 1977: *Spatial Population Analysis*. London: Edward Arnold.

REISER, R. 1973: The territorial illusion and behavioural sink: critical notes on behavioural geography. *Antipode* 5(3), 52–7.

RELPH, E. 1970: An inquiry into the relations between phenomenology and geography. *The Canadian Geographer* 14, 193–201.

RELPH, E. 1976: *Place and Placelessness*. London: Pion.

RELPH, E. 1977: Humanism, phenomenology, and geography. *Annals, Association of American Geographers* 67, 177–9.

REYNOLDS, R. B. 1956: Statistical methods in geographical research. *Geographical Review* 46, 129–32.

RIDDELL, J. B. 1970: *The Spatial Dynamics of Modernization in Sierra Leone*. Northwestern University Press, Evanston.

RIMMER, P. J. 1978: Redirections in transport geography. *Progress in Human Geography* 2, 76–100.

ROBINSON, A. H. 1956: The necessity of weighting values in correlation analysis of areal data. *Annals, Association of American Geographers* 46, 233–6.

ROBINSON, A. H. 1961: On perks and pokes. *Economic Geography* 37, 181–3.

ROBINSON, A. H. 1962: Mapping the correspondence of isarithmic maps. *Annals, Association of American Geographers* 52, 414–25.

ROBINSON, A. H. and BRYSON, R. A. 1957: A method for describing quantitatively the correspondence of geographical distributions. *Annals, Association of American Geographers* 47, 379–91.

ROBINSON, A. H., LINDBERG, J. B. and BRINKMAN, L. W. 1961: A correlation and regression analysis applied to rural farm densities in the Great Plains. *Annals, Association of American Geographers* 51, 211–21.

ROBSON, B. T. 1972: The corridors of geography. *Area* 4, 213–14.

ROBSON, B. T. and COOKE, R. U. 1976: Geography in the United Kingdom, 1972–1976. *Geographical Journal* 142, 3–72.

RODER, W. 1961: Attitudes and knowledge on the Topeka flood plain. In G. F. White (editor), *Papers on Flood Problems*, University of Chicago, Department of Geography, Research Paper 70, Chicago, 62–83.

RODGERS, A. 1955: Changing locational patterns in the Soviet pulp and paper industries. *Annals, Association of American Geographers* 45, 85–104.

ROSE, J. K. 1936: Corn yield and climate in the Corn Belt. *Geographical Review* 26, 88–102.

ROTHSTEIN, J. 1958: *Communication, Organization and Science*. Colorado: Falcon's Wing Press.

RUSHTON, G. 1969: Analysis of spatial behavior by revealed space preference. *Annals, Association of American Geographers* 59, 391–400.

SACK, R. D. 1972: Geography, geometry and explanation. *Annals, Association of American Geographers* 62, 61–78.

SACK, R. D. 1973a: Comment in reply. *Annals, Association of American Geographers* 63, 568–9.

SACK, R. D. 1973b: A concept of physical space in geography. *Geographical Analysis* 5, 16–34.

SACK, R. D. 1974a: The spatial separatist theme in geography. *Economic Geography* 50, 1–19.

SACK, R. D. 1974b: Chorology and spatial analysis. *Annals, Association of American Geographers* 64, 439–52.

SAMUELS, M. S. 1978: Existentialism and human geography. In D. Ley and M. S. Samuels (editors), *Humanistic Geography: Prospects and Problems*. Chicago: Maaroufa Press, 22–40.

SANTOS, M. 1974: Geography, Marxism and underdevelopment. *Antipode* 6(3), 1–9.

SAUER, C. O. 1925: The morphology of landscape, *University of California Publications in Geography* 2, 19–54.

SAUER, C. O. 1941: Foreword to historical geography. *Annals, Association of American Geographers* 31, 1–24.

SAUER, C. O. 1956: The education of a geographer. *Annals, Association of American Geographers* 46, 287–99.

SCHAEFER, F. K. 1953: Exceptionalism in geography: a methodological examination. *Annals, Association of American Geographers* 43, 226–49.

SHANNON, G. W. and DEVER, G. E. A. 1974: *Health Care Delivery: Spatial Perspectives*. New York: McGraw Hill.

SIDDALL, W. R. 1961: Two kinds of geography. *Economic Geography* 37, facing page 189.

SIMON, H. A. 1957: *Models of Man: Social and Rational*. New York: J. Wiley.

SINCLAIR, J. G. and KISSLING, C. C. 1971: A network analysis approach to fruit distribution planning. *Proceedings, Sixth New Zealand Geography Conference* Volume I, Christchurch, 131–6.

SLATER, D. 1973: Geography and underdevelopment—1. *Antipode* 5(3), 21–33.

SLATER, D. 1975: The poverty of modern geographical enquiry. *Pacific Viewpoint* 16, 159–76.

SMITH, C. T. 1965: Historical geography: current trends and prospects. In R. J. Chorley and P. Haggett (editors), *Frontiers in Geographical Teaching*, London: Methuen, 118–43.

SMITH, D. M. 1971: America! America? Views on a pot melting. 2. Radical geography—the next revolution? *Area* 3, 153–7.

SMITH, D. M. 1973a: Alternative 'relevant' professional roles. *Area* 5, 1–4.

SMITH, D. M. 1973b: *The Geography of Social Well-Being in the United States*. New York: McGraw Hill.

SMITH, D. M. 1977: *Human Geography: A Welfare Approach*. London: Edward Arnold.

SOJA, E. W. 1968: *The Geography of Modernization in Kenya*. Syracuse: Syracuse University Press.

SPATE, O. H. K. 1960a: Quantity and quality in geography. *Annals, Association of American Geographers* 50, 377–94.

SPATE, O. H. K. 1960b: Lord Kelvin rides again. *Economic Geography* 36, facing page 1.

SPATE, O. H. K. 1963: Letter to the Editor. *Geography* 48, 206.

SPENCER, H. 1892: *A System of Synthetic Philosophy, Volume I First Principles* (fourth edition), New York: Appleton.

STAMP, L. D. 1966: Ten years on. *Transactions, Institute of British Geographers* 40, 11–20.

STAMP, L. D. and BEAVER, S. H. 1947: *The British Isles*. London: Longman.

STEEL, R. W. 1974: The Third World: geography in practice. *Geography* 59, 189–207.

STEGMULLER, W. 1976: *The Structure and Dynamics of Theories*. New York: Springer-Verlag.

STEWART, J. Q. 1947: Empirical mathematical rules concerning the distribution and equilibrium of population. *Geographical Review* 37, 461–85.

STEWART, J. Q. 1956: The development of social physics. *American Journal of Physics* 18, 239–53.

STEWART, J. Q. and WARNTZ, W. 1958: Macrogeography and social science. *Geographical Review* 48, 167–84.

STEWART, J. Q. and WARNTZ, W. 1959: Physics of population distribution. *Journal of Regional Science* 1, 99–123.

STODDART, D. R. 1965: Geography and the ecological approach: the ecosystem as a geographic principle and method. *Geography* 50, 242–51.

STODDART, D. R. 1966: Darwin's impact on geography. *Annals, Association of American Geographers* 56, 683–98.

STODDART, D. R. 1967a: Growth and structure of geography. *Transactions, Institute of British Geographers* 41, 1–19.

STODDART, D. R. 1967b: Organism and ecosystem as geographic models. In R. J. Chorley and P. Haggett (editors), *Models in Geography*, London: Methuen, 511–48.

STODDART, D. R. 1975a: The RGS and the foundations of geography at Cambridge. *Geographical Journal* 141, 216–39.

STODDART, D. R. 1975b: Kropotkin, Réclus and 'relevant' geography. *Area* 7, 188–90.

STODDART, D. R. 1977: The paradigm concept and the history of geography. Abstract of a paper for the conference of the International Geographical Union Commission on the History of Geographic Thought, Edinburgh.

STOUFFER, S. A. 1940: Intervening opportunities: a theory relating mobility and distance. *American Sociological Review* 5, 845–67.

SUPPE, F. 1977a: The search for philosophic understanding of scientific theories. In F. Suppe (editor), *The Structure of Scientific Theories*, Urbana: University of Illinois Press, 3–233.

SUPPE, F. 1977b: Exemplars, theories and disciplinary matrices. In F. Suppe (editor), *The Structure of Scientific Theories*, Urbana: Univeristy of Illinois Press, 483–99.

SUPPE, F. 1977c: Afterword—1977. In F. Suppe (editor), *The Structure of Scientific Theories*, Urbana: University of Illinois Press, 617–730.

SVIATLOVSKY, E. E. and EELS, W. C. 1937: The centrographical method and regional analysis. *Geographical Review* 27, 240–54.

TAAFFE, E. J. 1974: The spatial view in context. *Annals, Association of American Geographers* 64, 1–16.

TAAFFE, E. J., MORRILL, R. L. and GOULD, P. R. 1963: Transport expansion in underdeveloped countries: a comparative analysis. *Geographical Review* 53, 503–29.

TATHAM, G. 1953: Environmentalism and possibilism. In G. Taylor (editor), *Geography in the Twentieth Century*, London: Methuen, 128–64.

TAYLOR, E. G. R. 1937: Whither geography? a review of some recent geographical texts. *Geographical Review* 27, 129–35.

TAYLOR, P. J. 1976: An interpretation of the quantification debate in British geography. *Transactions, Institute of British Geographers* NSI, 129–42.

TAYLOR, P. J. 1978: Political geography. *Progress in Human Geography*, 2, 153–62.

TAYLOR, P. J. and JOHNSTON, R. J. 1979: *Geography of Elections*. Harmondsworth: Penguin Books.

TAYLOR, P. J. and GUDGIN, G. 1976: A statistical theory of electoral redistricting. *Environment and Planning A* 8, 43–58.

THOMAN, R. S. 1965: Some comments on *The Science of Geography*. *The Professional Geographer* 17(6), 8–10.

THOMAS, E. N. 1960: Areal associations between population growth and selected factors in the Chicago Urbanized Area. *Economic Geography* 36, 158–70.

THOMAS, E. N. 1968: Maps of residuals from regression. In B. J. L. Berry and D. F. Marble (editors), *Spatial Analysis*, Englewood Cliffs: Prentice-Hall, 326–52.

THOMAS, E. N. and ANDERSON, D. L. 1965: Additional comments on weighting values in correlation analysis of areal data. *Annals, Association of American Geographers* 55, 492–505.

THOMAS, W. L. (editor) 1956: *Man's Role in Changing the Face of the Earth*. Chicago: University of Chicago Press.

THOMPSON, J. H. *et al.* 1962: Toward a geography of economic health: the case of New York state. *Annals, Association of American Geographers* 52, 1–20.

THRIFT, N. J. 1977: *An Introduction to Time Geography*. Concepts and Techniques in Modern Geography 13, Norwich: Geoabstracts Ltd.

THRIFT, N. J. 1979: Unemployment in the inner city: urban problem or structural imperative? a review of the British experience. In D. T. Herbert and R. J. Johnston (editors), *Geography and the Urban Environment: Progress in Research and Applications*, Volume 2. London: John Wiley, 125–226.

TIMMS, D. 1965: Quantitative techniques in urban social geography. In R. J. Chorley and P. Haggett (editors), *Frontiers in Geographical Teaching*, London: Methuen, 239–65.

TOULMIN, S. E. 1970: Does the distinction between normal and revolutionary science hold water? In I. Lakatos and A. Musgrave (editors), *Criticism and the Growth of Knowledge*, London: Cambridge University Press, 39–48.

TREWARTHA, G. T. 1973: Comments on geography and public policy. *The Professional Geographer* 25, 78–9.

TUAN, YI-FU 1971: Geography, phenomenology, and the study of human nature. *The Canadian Geographer* 15, 181–92.

TUAN, YI-FU 1974: Space and place: humanistic perspectives. In C. Board *et al.* (editors), *Progress in Geography* 6, London: Edward Arnold, 211–52.

TUAN, YI-FU 1975a: Images and mental maps. *Annals, Association of American Geographers* 65, 205–13.

TUAN, YI-FU 1975b: Place: an experiential perspective. *Geographical Review* 65, 151–65.

TUAN, YI-FU 1976: Humanistic geography. *Annals, Association of American Geographers* 66, 266–76.
TUAN, YI-FU 1978: Literature and geography: implications for geographical research. In D. Ley and M. S. Samuels (editors), *Humanistic Geography: Prospects and Problems*. Chicago: Maaroufa Press, 194–206.
TULLOCK, G. 1976: *The Vote Motive*. London: Institute of Economic Affairs.
ULLMAN, E. L. 1941: A theory of location for cities. *American Journal of Sociology* 46, 853–64.
ULLMAN, E. L. 1953: Human geography and area research. *Annals, Association of American Geographers* 43, 54–66.
ULLMAN, E. L. 1956: The role of transportation and the bases for interaction. In W. L. Thomas (editor), *Man's Role in Changing the Face of the Earth*, Chicago: University of Chicago Press, 862–80.
URLICH, D. U. 1972: Migrations of the North Island Maoris 1800–1840: a systems view of migration. *New Zealand Geographer* 28, 23–35.
URLICH CLOHER, D. 1975: A perspective on Australian urbanization. In J. M. Powell and M. Williams (editors), *Australian Space, Australian Time: Geographical Perspectives*, Melbourne: Oxford University Press, 104–59.
VAN DEN DAELE, W. and WEINGART, P. 1976: Resistance and receptivity of science to external direction: the emergence of new disciplines under the impact of science policy. In G. Lemaine *et al.* (editors), *Perspectives on the Emergence of Scientific Disciplines*, The Hague: Mouton, 247–75.
WAGNER, P. L. 1976: Reflections on a radical geography. *Antipode* 8(3), 83–5.
WALMSLEY, D. J. 1972: *Systems Theory: A Framework for Human Geographical Enquiry*. Research School of Pacific Studies, Department of Human Geography Publication HG/7, Australian National University, Canberra.
WALMSLEY, D. J. 1974: Positivism and phenomenology in human geography. *The Canadian Geographer* 18, 95–107.
WARD, D. 1971: *Cities and Immigrants: A Geography of Change in Nineteenth Century America*. New York: Oxford University Press.
WARNTZ, W. 1959a: *Toward a Geography of Price*. Philadelphia: University of Pennsylvania Press.
WARNTZ, W. 1959b: Geography at mid-twentieth century. *World Politics* 11, 442–54.
WARNTZ, W. 1959c: Progress in economic geography. In P. E. James (editor), *New Viewpoints in Geography*, Washington: National Council for the Social Studies, 54–75.
WARNTZ, W. 1968: Letter to the Editor. *The Professional Geographer* 20, 357.

WATKINS, J. W. N. 1970: Against 'normal science'. In I. Lakatos and A. Musgrave (editors), *Criticism and the Growth of Knowledge*, London: Cambridge University Press, 25–38.

WATSON, J. D. 1968: *The Double Helix: A Personal Account of the Discovery of the Structure of DNA*. London: Weidenfeld and Nicholson.

WATSON, J. W. 1953: The sociological aspects of geography. In G. Taylor (editor), *Geography in the Twentieth Century*, London: Methuen, 463–99.

WATSON, J. W. 1955: Geography: a discipline in distance. *Scottish Geographical Magazine* 71, 1–13.

WATTS, S. J. and WATTS, S. J. 1978: The idealist alternative in geography and history. *The Professional Geographer* 30, 123–7.

WEAVER, J. C. 1943: Climatic relations of American barley production. *Geographical Review* 33, 569–88.

WEAVER, J. C. 1954: Crop-combination regions in the Middle West. *Geographical Review* 44, 175–200.

WEBBER, M. J. 1972: *Impact of Uncertainty on Location*. Canberra: Australian National University Press.

WEBBER, M. J. 1977: Pedagogy again: what is entropy? *Annals, Association of American Geographers* 67, 254–66.

WESTERN, J. S. 1978: Knowing one's place: 'the Coloured people' and the Group Areas Act in Cape Town. In D. Ley and M. S. Samuels (editors), *Humanistic Geography: Prospects and Problems*. Chicago: Maaroufa Press, 297–318.

WHITE, G. F. 1945: *Human Adjustment to Floods*. University of Chicago, Department of Geography, Research Paper 29, Chicago.

WHITE, G. F. 1972: Geography and public policy. *The Professional Geographer* 24, 101–4.

WHITE, G. F. 1973: Natural hazards research. In R. J. Chorley (editor), *Directions in Geography*, London: Methuen, 193–216.

WHITEHAND, J. W. R. 1970: Innovation diffusion in an academic discipline: the case of the 'new' geography. *Area* 2(3), 19–30.

WHITEHAND, J. W. R. and PATTEN, J. H. C. (editors) 1977: *Change in the Town*. Transactions, Institute of British Geographers 2(3)

WILLIAMS, P. R. 1978: Urban managerialism: a concept of relevance? *Area* 10, 236–40.

WILSON, A. G. 1967: A statistical theory of spatial distribution models. *Transportation Research* 1, 253–69.

WILSON, A. G. 1970: *Entropy in Urban and Regional Modelling*. London: Pion Ltd.

WILSON, A. G. 1974: *Urban and Regional Models in Geography and Planning*. London: John Wiley.

WILSON, A. G. 1976a: Catastrophe theory and urban modelling: an application to modal choice. *Environment and Planning A* 8, 351–46.

WILSON, A. G. 1976b: Retailers' profits and consumers' welfare in a spatial interaction shopping model. In I. Masser (editor), *Theory and Practice in Regional Science*, London: Pion Ltd., 42–57.

WILSON, A. G. 1978: Mathematical education for geographers. Department of Geography, University of Leeds, *Discussion Paper 211*, Leeds.

WILSON, A. G., REES, P. H. and LEIGH, C. 1977: *Models of Cities and Regions*. London: John Wiley.

WISE, M. J. 1977: On progress and geography. *Progress in Human Geography* 1, 1–11.

WISNER, B. 1970: Introduction: on radical methodology, *Antipode* 2, 1–3.

WOLDENBERG, M. J. and BERRY, B. J. L. 1967: Rivers and central places: analogous systems? *Journal of Regional Science* 7, 129–40.

WOLF, L. G. 1976: Comments on the Harries-Peet controversy. *The Professional Geographer* 28, 196–8.

WOLPERT, J. 1964: The decision process in spatial context. *Annals, Association of American Geographers* 54, 337–58.

WOLPERT, J. 1965: Behavioral aspects of the decision to migrate. *Papers and Proceedings, Regional Science Association* 15, 159–72.

WOLPERT, J. 1967: Distance and directional bias in inter-urban migratory streams. *Annals, Association of American Geographers* 57, 605–16.

WOLPERT, J. 1970: Departures from the usual environment in locational analysis. *Annals, Association of American Geographers* 60, 220–9.

WOLPERT, J., DEAR, M. and CRAWFORD, R. 1975: Satellite mental health facilities. *Annals, Association of American Geographers* 65, 24–35.

WOOLDRIDGE, S. W. 1956: *The Geographer as Scientist*. London: Thomas Nelson.

WOOLDRIDGE, S. W. and EAST, W. G. 1958: *The Spirit and Purpose of Geography*. London: Hutchinson.

WOOLGAR, S. W. 1976: The identification and definition of scientific collectivities. In G. Lemaine *et al.* (editors), *Perspectives on the Emergence of Scientific Disciplines*, The Hague: Mouton, 233–45.

WRIGHT, J. K. 1925: *The Geographical Lore of the Time of the Crusades: A Study in the History of Medieval Science and Tradition in Western Europe*. New York: American Geographical Society.

WRIGHT, J. K. 1947: *Terrae incognitae*: the place of imagination in geography. *Annals, Association of American Geographers* 37, 1–15.

WRIGLEY, E. A. 1965: Changes in the philosophy of geography. In R. J. Chorley and P. Haggett (editors), *Frontiers in Geographical Teaching*, London: Methuen, 3–24.

ZELINSKY, W. 1970: Beyond the exponentials: the role of geography in the great transition. *Economic Geography* 46, 499–535.

ZELINSKY, W. 1973a: Women in geography: a brief factual report. *The Professional Geographer* 25, 151–65.

ZELINSKY, W. 1973b: *The Cultural Geography of the United States.* Englewood Cliffs: Prentice-Hall.

ZELINSKY, W. 1974: Selfward bound? Personal preference patterns and the changing map of American society. *Economic Geography* 50, 144–79.

ZELINSKY, W. 1975: The demigod's dilemma. *Annals, Association of American Geographers* 65, 123–43.

ZIPF, G. K. 1949: *Human Behavior and the Principle of Least Effort.* New York: Hafner.

Additional bibliography, January 1981

Since this book was completed in late 1978, many publications have appeared which are relevant to its major themes. Only the most important are listed here, as major contributions to the literature on the history of human geography. Certain journals—notably *Progress in Human Geography* with its annual Progress Reports on various topics—contain a lot of relevant material, but this is not listed here. Two special issues of journals deserve particular mention, however. The October 1978 (Volume 22, No. 1) of *American Behavioural Scientist* is entitled 'Human Geography: Coming of Age' and contains several valuable review papers (edited by Wilbur Zelinsky). The March 1979 (Volume 19, No. 1) of *Annals, Association of American Geographers* contains a number of personal recollections of the history of geography in the U.S.A.

BENNETT, R. J. and WRIGLEY, N. (editors) 1981: *Quantitative Geography in Britain: Retrospect and Prospect.* London: Routledge and Kegan Paul.

BERRY, B. J. L. (editor) 1978: *The Nature of Change in Geographical Ideas.* De Kalb: Northern Illinois University Press.

BLOUET, B. W. and LAWSON, M. P. (editors) 1981: *The Origins of Academic Geography in the United States.* Boston: Archon Press.

BROWN, E. H. (editor) 1980: *Geography Yesterday and Tomorrow.* Oxford: Oxford University Press.

FREEMAN, T. W. 1980: *A History of Geography in Great Britain.* London: Longman.

GALE, S. and OLSSON, G. (editors) 1978: *Philosophy in Geography.* The Hague: D. Reidel.

HARVEY, M. E. and HOLLY, B. P. (editors) 1981: *Themes in Geographic Thought.* London: Croom Helm.

HUGGETT, R. J. 1980: *Systems Analysis in Geography.* Oxford: Oxford University Press.

JAMES, P. E. and MARTIN, G. J. 1979: *The Association of American Geographers: The First Seventy-Five Years*. Washington D.C.: The Association of American Geographers.

JOHNSTON, R. J. *et al*. (editors) 1981: *Dictionary of Human Geography*. Oxford: Blackwell.

SACK, R. D. 1980: *Conceptions of Space in Social Thought*. London: Macmillan.

Index

General Index

Index to Authors